INSECTS, SCIENCE, & SOCIETY

ACADEMIC PRESS RAPID MANUSCRIPT REPRODUCTION

Proceedings of a Symposium
on Insects, Science, and Society
Held at Cornell University, Ithaca
New York, October 14–15, 1974

INSECTS, SCIENCE, & SOCIETY

Edited by

David Pimentel

Department of Entomology
Cornell University
Ithaca, New York

Academic Press, Inc.

New York San Francisco London 1975

A Subsidiary of Harcourt Brace Jovanovich, Publishers

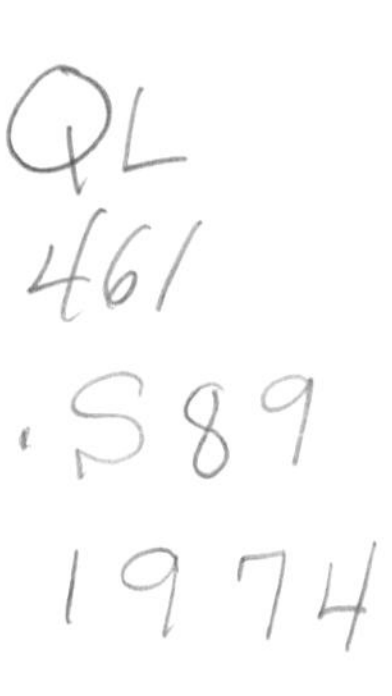

ACADEMIC PRESS, INC.
111 Fifth Avenue, New York, New York 10003

United Kingdom Edition published by
ACADEMIC PRESS, INC. (LONDON) LTD.
24/28 Oval Road, London NW1

Library of Congress Cataloging in Publication Data

Symposium on Insects, Science, and Society, Cornell University, 1974.
Insects, science, and society.

Includes bibliographical references and index.
1. Insects–Congresses. 2. Insect control–Congresses. I. Pimentel, David, (date) II. Title.
QL461.S89 1974 595.7'05 75-13079
ISBN 0-12-556550-X

PRINTED IN THE UNITED STATES OF AMERICA

To John Henry Comstock

CONTENTS

IV. INSECT POPULATION DYNAMICS

V. INSECT-PEST MANAGEMENT

CONTRIBUTORS

Richard D. Alexander, Museum of Zoology and Department of Zoology, The University of Michigan, Ann Arbor, Michigan 48104

Howard E. Evans, Department of Zoology and Entomology, Colorado State University, Fort Collins, Colorado 80521

John S. Kennedy, ARC Insect Physiology Group, Imperial College Field Station, Silwood Park, Ascot, Berks SL5 7PY, United Kingdom

Waldemar Klassen, National Program Staff, Agricultural Research Service, U.S. Department of Agriculture, Beltsville, Maryland 20705

John J. McKelvey, Jr., The Rockefeller Foundation, 111 West 50th Street, New York, New York 10020

Powers S. Messenger, Department of Entomological Sciences, Agricultural Experiment Station, 137 Giannini Hall, University of California, Berkeley, California 94720

L. Dale Newsom, Department of Entomology, Agricultural Experiment Station, Center for Agricultural Science and Rural Development, Louisiana State University, Baton Rouge, Louisiana 70803

Mano D. Pathak, The International Rice Research Institute, Los Baños, Laguna, Philippines

David Pimentel, Department of Entomology, Comstock Hall, Cornell University, Ithaca, New York 14853

Wendell L. Roelofs, New York State Agricultural Experiment Station, Geneva, New York 14456

Edward H. Smith, Department of Entomology, Comstock Hall, Cornell University, Ithaca, New York 14853

T. R. E. Southwood, Department of Zoology and Applied Entomology, Imperial College, London, United Kingdom

Edward O. Wilson, Museum of Comparative Zoology, Harvard University, Cambridge, Massachusetts 02138

FOREWORD

Centennial of Entomology

Cornell University celebrated its centennial in 1968 and now, some six years later, we celebrate Cornell's centennial of entomology. This demonstrates that the discipline of entomology was recognized early as a building block of a great university. In the development of an institution, it is usually the case that the vision of a single intellect provides the initial incentive and leadership around which other individuals build. In entomology at Cornell, that initiator was John Henry Comstock. This centennial celebration is essentially a recognition of Comstock, and Prof. Howard Evans in "*The Comstock Heritage*" provides an historical perspective of the man and his work. His vision and zeal led to the organization of a department of entomology, the first in the United States. His intellectual effort established the biological framework for the orderly study of entomology, and the prolificness of his pen provided the most widely used text in entomology. His quality of sharing attracted around him a cluster of gifted individuals who added substance to the department and to the Comstock vision.

Comstock's concern was for mankind. He considered medicine and the ministry and concluded finally that he could best serve in natural history, specifically entomology. It is appropriate, therefore, to honor the memory of John Henry Comstock with a symposium addressing itself broadly to the needs of society and the role which the science of entomology can play in meeting these needs. As we ponder the problems facing mankind today, we appreciate more fully the prophetic vision of Comstock regarding the interaction between man and insects.

This symposium and the resulting book on *Insects, Science, and Society*, edited by Prof. David Pimentel, is appropriately dedicated to John Henry Comstock.

Edward H. Smith

PREFACE

Insects, Science, and Society is a collection of papers presented at a symposium of the same name celebrating the centennial of entomology at Cornell University. The symposium and the resulting book are unique in having assembled some of the world's leading entomologists, who discussed recent advances in their diverse specialties. The value of this volume rests with the outstanding contributors.

In commemorating the hundred years of entomology, we also honor the founder of the department at Cornell, John Henry Comstock, whose investigations encompassed not only many aspects of entomology—insect taxonomy, morphology, and ecology—but also focused on practical problems of pest control. His concern was with how the study of insects relates to other of the biological sciences and, further, how it benefits human society. We thus felt "Insects, Science, and Society" would be an appropriate framework for this symposium.

Prof. Edward H. Smith and Howard E. Evans provide an historical perspective of entomology and, in particular, John Henry Comstock, setting the stage for the other outstanding papers discussing aspects of the social implications of insects, communication among insects, other interactions among insects and between insects and plants, population dynamics, and pest management.

These presentations are exceptional from the standpoints of scholarship and relevance to current knowledge; the book as a whole has much to offer students, teachers, and researchers.

ACKNOWLEDGMENTS

A book such as this is the product of the efforts and cooperation of a great many people. First, I want to thank all the contributors, not only for their papers, but also for their prompt help in dealing with minor editorial-production problems. My sincere thanks are expressed to Ms. Lynn Colvin for her editorial assistance. Thanks are due to Ms. Maxine Dattner and Ms. Robin Wolaner for their part in preparing the manuscripts. I am grateful to the staff of Academic Press for their interest in our project and for expediting the publication of the symposium papers. Last and most important, I want to acknowledge the financial support provided in part by Ford Foundation's program on the Ecology of Pest Management and the Grace H. Griswold Fund that helped make the symposium a success.

David Pimentel

THE COMSTOCK HERITAGE

Howard E. Evans

In 1928 L.O. Howard, one of Comstock's most distinguished students, commented on his teacher in the following words (Howard 1930):

> Who shall say that in the future, when the vital importance of insects as affecting the well-being of humanity shall have become fully realized, this spot shall not become in a way a shrine where entomologists will gather in token of their respect to the first great teacher of entomology in America?

Almost half a century later, these words seem prophetic. We appear to be on the verge of a world food shortage of major proportions, and the role of insects as competitors of man will soon be fully appreciated. The department Comstock founded in 1874 has figured significantly in the growth of entomology, especially through its many students now scattered all over the world, engaged in research on many topics and in turn training their own students.

It is easy to forget that only a century ago entomology scarcely existed as a discipline. W.D. Peck taught entomology at Harvard on a somewhat irregular basis in the 1830s. Thomas Say at one time held a professorship at the University of Pennsylvania and is said to have done some teaching on entomology. A.J. Cook, at Michigan Agricultural College, began teaching entomology about 1867, although he was actually a professor of mathematics. At Kansas Agricultural College, B.F. Mudge began lecturing on economic entomology about 1870. T.W. Harris's influential book *Report on Insects Injurious to Vegetation* had been published in 1841, and A.S. Packard, a student of Louis Agassiz, published a major text, *A Guide to the Study of Insects*, in 1869. C.V. Riley's historic *Missouri Reports* began in 1868. So we can say that by the 1870s, when Comstock was attending Cornell, the stage was set for the emergence of entomology as a discrete discipline.

John Henry Comstock was in every sense a self-made man. His father died when he was an infant, and his mother worked as a nurse and later remarried, so that mother and son saw very little of each other. He spent some time in an orphan asylum and had several different foster parents. For some years he served as a cook on various ships on the Great Lakes. His early education was spotty, consisting mostly of books he picked up here and there. From 1867 to 1869 he attended Falley Seminary at Fulton, New York. While considering college, he heard a visiting clergyman comment on "that godless institution, Cornell University." This pricked Henry's curiosity to the extent that he obtained a catalog and learned that Agassiz was to teach there, and that a wide variety of scientific courses were being offered.

When John Henry Comstock arrived at Cornell in the fall of 1869, he had read Harris's *Insects Injurious to Vegetation* and was already filled with enthusiasm for insects, although at the time he was considering medicine as a career. As a sophomore in 1872, Comstock gave a series of lectures in entomology, resulting from a petition submitted to the dean by thirteen students, who made up his first class. Comstock was appointed instructor of entomology in 1873, although he did not graduate until 1874. His salary for the first year was $500, which, he commented, was "enough to support me nicely."

Comstock's marriage to Anna Botsford in 1878 was a milestone in his life. There can be nothing more important to a scientist than having a wife who not only understands and appreciates his work, but is willing to help him in every way possible. She undertook the training needed to make her an outstanding scientific illustrator, and her fine engravings are to be found in virtually all of her husband's major publications. Later she developed the Department of Nature Study at Cornell and was as much of a pioneer in this field as her husband was in entomology. Indeed, her bibliography rivals that of her husband, and her *Handbook of Nature Study* went through twenty-four editions and is still in print and a classic in its field.

All of this has been told very well by Mrs. Comstock in her book *The Comstocks of Cornell*, and there seems no need to go over the details of their lives at great length (Comstock 1953). They overcame a great many handicaps in achieving their many successes. As a sophomore at Cornell, Henry Comstock was totally without funds and remarked that he probably would have left school except that he did not even have enough money to leave town. He worked on the construction of McGraw Hall, and later as chimesmaster. Even as a federal entomologist in Washington, a position he held from 1879 through 1881, he was often impecunious. L.O. Howard, who was his assistant, remarks that when they wanted to go to the theater, Howard had to pawn his watch—Comstock did not even own one (Howard 1930).

Comstock was plagued by ill health. He had lung problems and at one time was suspected of having tuberculosis; he had a serious attack of typhoid fever; and for years he suffered periodic attacks of malaria. He had a nervous temperament, resulting in short fits of temper and occasional relapses into the stuttering of his youth. When he first entered Cornell, Burt Wilder, whose assistant he was, described him as "an inspired anthropomorphic squirrel," and a few years later Mrs. Comstock found him to be "very active [moving] with a great rapidity that gave the observer a breathless sensation" (Comstock 1953).

Despite idiosyncrasies, Comstock had all the attributes one expects to find in a successful scientist: overwhelming enthusiasm for his subject; concern for his students, his colleagues, and the needs of the layman; and tremendous self-discipline. Much of his research and writing was done between 4 A.M. and breakfast, the only period he could find that was free from other duties and responsibilities after he became a professor and department head.

A brief look at some of Comstock's major publications may help us to understand his thinking and the reasons for his influence on students and on entomology. His first major work, resulting from a trip to the southern states in 1878 to study cotton insects, was published in 1880. It was a carefully documented report which was said to be well received by farmers as well as by Comstock's scientific peers. Even Charles Darwin wrote and thanked him, saying that he found many interesting facts bearing on his theory of evolution. The report deals mostly with the cotton leaf worm, *Alabama argillacea*, and the bollworm, *Heliothis zea*, and deals at length with the effect of weather conditions on populations and with natural and artificial controls of these insects.

During his tenure as a federal entomologist, Comstock became immersed in the study of scale insects and wrote three papers (1880, 1882, and 1883) dealing mostly with the systematics of Coccidae. These pioneering studies of American scale insects were of such importance that they were reprinted without change as a Cornell Experiment Station Bulletin in 1916. Comstock described many new species, and was the first to discover the San José scale, which he called *Aspidiotus perniciosus*. One of his students, Alexander MacGillivray, later renamed the genus for his mentor, and the insect became *Comstockaspis perniciosus*. In the course of time he had many insects named for him, including the Comstock mealybug, *Pseudococcus comstocki*, a serious pest of citrus. I am sure only a dedicated entomologist would appreciate having a scale insect or a mealybug named after him.

Comstock's lecture syllabus was published in 1888 under the title *An Introduction to Entomology*, and in 1895 an expanded version, entitled *A Manual for the Study of Insects*, appeared. The first of these was privately published, and the second marked the permanent establishment of the Comstock Publishing Company. In 1869 Cornell had set up the first university press in America, but by 1884 it had passed out of existence. Although Comstock was told that his publishing enterprise had little chance of success, in fact sales boomed. Later on Anna Comstock's *Handbook of Nature Study* sold hundreds of thousands of copies and was the mainstay of the press, which eventually became an important component of a revitalized university press. When Comstock died in 1931 he left not only the publishing company to the university, but also his home and offices, in which the press has since been located (Bishop 1962).

The Comstock Publishing Company's first venture was *The Wilder Quarter-Century Book*, a memorial to Comstock's teacher, Burt G. Wilder, on the completion of twenty-five years of service to Cornell. Comstock's contribution was entitled "Evolution and Taxonomy: An Essay on the Application of the Theory of Natural Selection in the Classification of Animals and Plants, Illustrated by a Study of the Wings of Insects and by a Contribution to the Classification of the Lepidoptera." This proved to be one of its author's most important papers and is worth quoting in part (bear in mind that these words were written in 1893):

> It is now 34 years since the publication of Darwin's Origin of Species; and the great war of opinions . . . has been fought to a conclusion. . . . Notwithstanding this I do not believe that the systematists of to-day are making as much use of the theory of descent in taxonomic work as they might. We are still busy describing species as if they were immutable entities; and in our descriptions we give little thought to the causes that have determined the forms of organisms. . . . Here I believe lies the work of the systematist of the future.

Comstock follows this with a fairly lengthy discussion of the importance of determining the function and adaptive value of various organs. He also discusses the principle of irreversibility; that is, as he says, "When an organ has wholly disappeared in a genus, other genera which originate as offshoots from this genus cannot regain the organ, although they might develop a substitute for it." This is nowadays called "Dollo's law," although in fact several people had expressed it before Dollo (Comstock attributed it to Meyrick). Comstock then applied the principles he had outlined to a study of insect wings, especially those of Lepidoptera. It was in this paper that he first stated his belief that there were homologies in the wing venation of all insects. He adopted the system of Redtenbacher, using names such as costa, subcosta, radius, and so forth, with which all entomologists are familiar. It was here, too, that he first separated the Lepidoptera into two suborders, Jugatae and Frenatae, according to the way the front and hind wings were held together. By no means all of Comstock's conclusions are still accepted; after all, that was eighty-one years ago. The names Jugatae and Frenatae still persist as the names of the entomology seminars at Cornell and the University of Minnesota, respectively, but the classification of Lepidoptera has undergone many changes through the years.

Comstock is perhaps most widely known for his system of wing venation, later developed in greater detail with J.G. Needham and usually referred to as the "Comstock-Needham system." This was first outlined in a series of articles in *The American Naturalist*, published in 1898 and 1899, and later elaborated still further in Comstock's book *The Wings of Insects*, published in 1918. In the words of British entomologist A.D. Imms (1931), "We owe to Comstock and Needham what is one of the greatest advances ever made in the study of insects."

The Comstocks were much concerned with interesting the layman in entomology and in nature study. In 1897 they produced a popular book called *Insect Life*, and in 1904 *How to Know the Butterflies*. With one of his students, Vernon L. Kellogg, Comstock produced a technical treatise called *The Elements of Insect Anatomy* (1897).

Comstock had long been interested in spiders, and around 1900 he began an intensive study of this group of arthropods. This resulted in one of his most important contributions, *The Spider Book* (1912). Like most of Comstock's books, this went through several editions, and in the 1948 reprinting Willis Gertsch had this to say in the Preface:

> Appearing at a time when arachnology was the property of a few trained specialists [*The Spider Book*] opened the way to a new appreciation of spiders and their near relatives. . . . It brought together for the first time in concise

> form a wealth of information on the structure, habits, and classification of the American arachnids. . . . No comparable work has appeared in any other language.

In preparing this book, Comstock developed some new techniques: he placed live spiders in boxes with a dark background, so that they would spin fresh webs that could then be photographed. He called his boxes "looms" and the webs "made-to-order webs." The photographs of webs that fill *The Spider Book* were unique in their time.

Finally, ten years after his retirement, the first of several editions of *An Introduction to Entomology* appeared (1924). This was a vastly more comprehensive and scholarly treatise on insects than his 1888 book of the same title. Many persons of my generation were indeed introduced to entomology through Comstock's book, and even today there is no American entomology text with anything like its comprehensive coverage.

As his publications demonstrate, during the latter part of his career Comstock was drawn to the basic aspects of entomology: systematics, evolution, morphology, and life-history studies. This was not the case earlier in his career. Andrew D. White, Cornell's first president, had urged him to keep in mind the interests of the farmer, and at this time Comstock wrote for various agricultural journals and talked regularly at the meetings of the New York Horticultural Society. But Comstock, despite his great energy, could not be all things to all people. In 1890 he was fortunate to have a brilliant student with a penchant for practical matters, Mark Slingerland. Slingerland became an instructor and later an assistant professor, producing a series of beautifully illustrated bulletins and quickly earning the respect of farmers and colleagues.

A few years later Alexander MacGillivray, a specialist in taxonomy, joined the staff, followed shortly thereafter by W.A. Riley, a medical entomologist. None of these men remained on the staff permanently, MacGillivray moving later to the University of Illinois and Riley to Minnesota. Slingerland died an untimely death in 1909 and was replaced by Glenn W. Herrick. In the meantime James G. Needham had joined the staff to develop the field of limnology, and C.R. Crosby, to broaden the program in economic entomology. By the time of Comstock's retirement in 1914, J.C. Bradley, Robert Matheson, J.T. Lloyd, G.C. Embody, and O.A. Johannsen had joined the staff, bringing the total to nine (Needham 1946). The age of specialization was well advanced, and the entomology department was well on its way to an expansion that could scarcely have been imagined in 1872, when Comstock was first asked to give a series of lectures on insects.

To list all of Comstock's students who attained distinction would take much too long. The list would include several persons already mentioned, such as L.O. Howard, Vernon Kellogg, Mark Slingerland, G.W. Herrick, J.G. Needham, C.R. Crosby, J.C. Bradley, W.A. Riley, and others. It would also include R.N. Chapman, the Minnesota ecologist; E.P. Felt, for many years state entomologist of New York; A.P. Morse, the orthopterist; G.C. Crampton, the morphologist; Nathan Banks, for many years curator of insects at Harvard; S.W. Frost, of Pennsylvania State University; and many others who studied under Comstock for at least brief periods. In a sense, Comstock was teacher to several generations of entomologists, through the influence of his books and through the influence of his students and his students' students.

What, then, is the Comstock heritage? It is more than Comstock Hall, the Comstock Memorial Library, the Comstock Press, and even more than the department he founded. It is a sense of the unity of entomology, a conviction of the interdependence of all its parts: systematics, morphology, physiology, life histories and behavior, and control. It is a sense of the interdependence of entomology with such other disciplines as botany, zoology, pathology, and chemistry. It is a sense of growth and innovation, of meeting new challenges with new approaches and new techniques. It is a feeling for sound scholarly work; Comstock's most quoted remark is his adage: "Be sure you are right, and then look again." Comstock was above all a synthesizer and an eclectic. His view of entomology was a broad one, embracing the best research of his time from all parts of the world.

What would Comstock think about this symposium? He would be delighted, but perhaps not greatly surprised, to find that sound taxonomy is more than ever essential to control, especially to biological control; to find that morphology, particularly the structure of exocrine glands and receptors, has proved of prime importance; to find that insect control is being studied in a broad context of ecology, behavior, and physiology; to find that insects' hormones, pheromones, and behavioral responses are being turned against them. He would be pleased, too, at the major contributions the study of insects has made to basic science: to genetics, evolution, behavior, and ecology. Comstock was above all concerned with insects as living organisms, such unique and fascinating organisms that those who study them are perhaps the most fortunate people on earth.

REFERENCES

Bishop, M. 1962. A history of Cornell. Cornell Univ. Press, Ithaca.

Comstock, A.B. 1953. The Comstocks of Cornell: John Henry Comstock and Anna Botsford Comstock. Edited by G.W. Herrick and R.G. Smith. Comstock Publ. Assoc., Ithaca.

Comstock, J.H. 1879. Report upon cotton insects. Special Report, U.S. Department of Agriculture. Govt. Printing Office, Washington.

———. 1881. Report of the entomologist. Part II. Report on scale insects, p. 276-349. *In* The annual report of the Department of Agriculture for the year 1880. Govt. Printing Office, Washington.

———. 1882. Report of the entomologist, p. 209-14. *In* Report of the Department of Agriculture for 1881 and 1882. Govt. Printing Office, Washington.

———. 1883. Second report on scale insects, p. 47-143. *In* Second report of the Department of Entomology of the Cornell University Experiment Station. Andrus & Church, Ithaca.

———. 1916. Reports on scale insects. Cornell Univ. Agr. Exp. Sta. Bull. 372:421-603.

———. 1888. An introduction to entomology. The author, Ithaca.

———. 1893. Evolution and taxonomy, an essay on the application of the theory of natural selection in the classification of animals and plants, illustrated by a study of the wings of insects and by a contribution to the classification of the Lepidoptera, p. 34-114. *In* The Wilder quarter-century book. Comstock Publ. Co., Ithaca.

———. 1897. Insect life. Appleton & Co., New York.

———. 1912. The spider book. Doubleday, Page & Co., Garden City.

———. 1918. The wings of insects. Comstock Publ. Co., Ithaca.

———. 1924. An introduction to entomology. Comstock Publ. Co., Ithaca.

Comstock, J.H., and A.B. Comstock. 1895. A manual for the

study of insects. Comstock Publ. Co., Ithaca.
———. 1904. How to know the butterflies. Appleton & Co., New York.
Comstock, J.H., and V.L. Kellogg. 1897. The elements of insect anatomy. Comstock Publ. Co., Ithaca.
Gertsch, W.J. 1948. Preface to The spider book, by J.H. Comstock. Revised and edited by W.J. Gertsch. Comstock Publ. Assoc., Ithaca.
Howard, L.O. 1930. A history of applied entomology (somewhat anecdotal). Smithsonian Miscellaneous Collections, vol. 84. Smithsonian Institution, Washington.
Imms, A.D. 1931. Recent advances in entomology. Blakiston's Son & Co., Philadelphia.
Needham, J.G. 1946. The lengthened shadow of a man and his wife. I and II. Sci. Monthly 62:140-50, 219-29.

INSECTS, SCIENCE, & SOCIETY

INTRODUCTION

David Pimentel

Insects, science, and society are intimately interrelated. Insects, the most abundant animal species, inhabit all corners of the earth and intrude as both benefactor and pest in man's endeavors. Insects also interact with other animal species, and with plants as well. Thus it is not strange that many biologists have turned their scientific focus on the study of insects. In the next critical twenty-five years entomology will play a special role in the well-being of man. Four major challenges face mankind during this final quarter of the twentieth century. These are human overpopulation, food shortages, fuel shortages, and environmental degradation.

In 1975 the world population is expected to reach four billion humans (National Academy of Sciences 1971). Based upon current growth rates and even allowing for reductions in birthrates in several countries, a study committee of the National Academy of Sciences estimates that the world population will reach at least seven billion by the year 2000 (fig. 1). The committee concludes that nothing short of a major catastrophe, such as a nuclear war, can stop this explosive increase (NAS 1971).

This rapid population growth corresponds to the exponentially increased use of fossil fuel by man (fig. 1). Fossil energy is used not only to improve the quality of life, but also to control disease and to improve agricultural production to feed the world population, thereby contributing significantly to the rapid population growth (NAS 1971).

Unfortunately, reducing death rates by effective disease control can almost instantaneously increase population growth rates (Freedman and Berelson 1974). For example, in Ceylon (in 1946-47) after the spraying with DDT to control malaria-carrying mosquitoes, the death rate in the human population fell in one year from 20 to 14 per thousand (Political and Economic Planning 1955). The resulting rapid population growth might have been averted if population controls had been initiated along with the public-health programs. Similarly, after DDT was used in Mauritius, death rates fell in one year from 27 to 15 per thousand, and in two years the

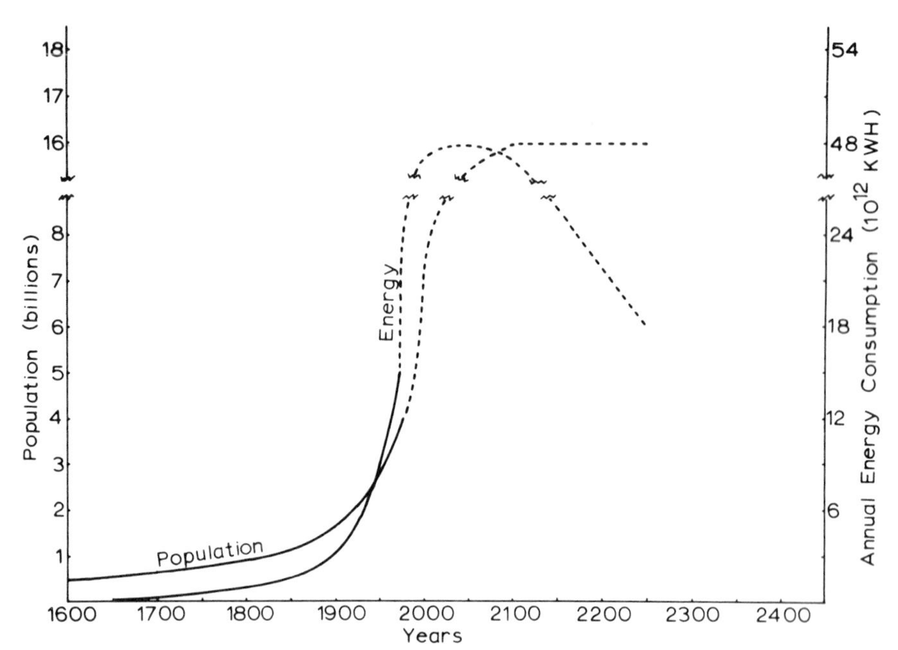

Fig. 1. Estimated world population numbers (———) from 3000 B.C. to 1973 A.D. and projected numbers (......) to 2500 A.D. (NAS 1971). Estimated fossil-fuel consumption (- - - -) from 1850 A.D. to 1973 A.D. and projected consumption (—·—·) to the year 2500 (data from Hubbert 1962).

population growth rate increased 7-fold—from about 5 to 35 per thousand (fig. 2).

The increase in human numbers has obviously resulted in a greater need for food supplies. The developed nations have responded to this need by sending surplus food, agricultural advisors, and farm supplies and equipment to the areas where the food and supplies are needed. Unfortunately, experience has shown that increasing food supplies is "painfully slow" (Brown 1965). Because many of the people in less developed countries are involved in agriculture, cultural and social habits of the people also must be modified if food production is to be increased (NAS 1971). Furthermore, the successful implementation of new crop-production technology is an expensive and complex endeavor.

For these and other reasons, world food production is lagging behind population growth (Food and Agriculture

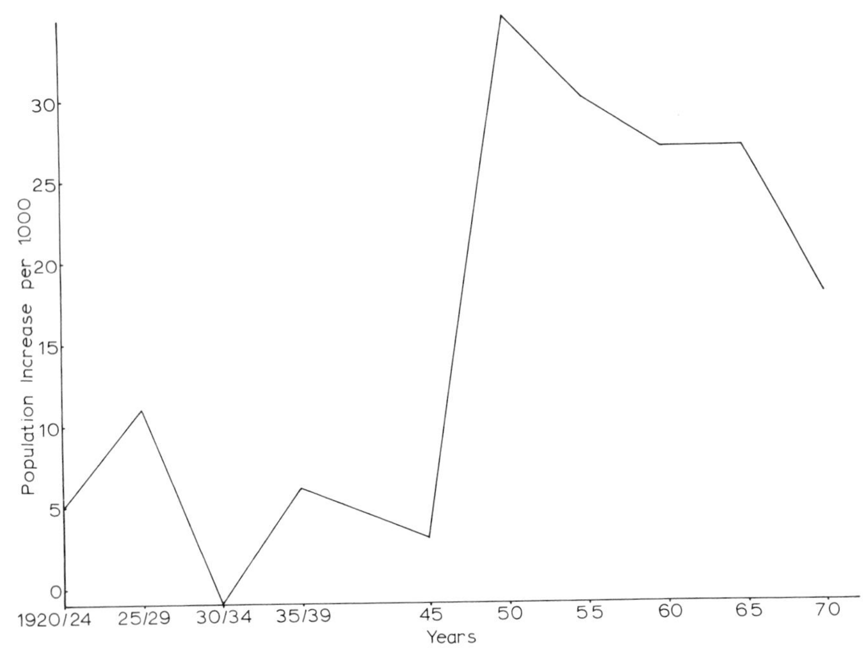

Fig. 2. Population growth rate on Mauritius from 1920 to 1970. Note from 1920 to 1945 the growth rate was about 5 per thousand, whereas after malaria control in 1945 the growth rate exploded to about 35 per thousand and has since very slowly declined (PEP 1955; United Nations 1957-1971). After 25 years the rate of increase is still nearly 4 times the 1920-45 level.

Organization 1970; NAS 1971). Today, with a population of about four billion humans, two-thirds of the world population is living on an average of about 2100 kilocalories per day, and most of these people are considered malnourished in protein-calorie consumption (President's Science Advisory Committee 1967; Spengler 1968; Woodham 1971; Borgstrom 1973). Sen (1965) remarked that those "half-fed" are in reality "half-alive"!

Although relatively few people in the United States are malnourished (the average kilocalorie consumption is about 3300 per day), increased demand for food at home and abroad has resulted in an increase in food prices of about 20% in 1973 (U.S. Department of Labor 1973) and probably the same in 1974. The food problem in the United States, however, seems minor compared with that of the remainder of the world.

Food is in short supply because of increasing human numbers. Energy is in short supply primarily for the same reason. The interrelationship between energy and human numbers is interesting, as well as alarming. Energy use has been increasing faster than the world population. Whereas the U.S. population doubled in the past sixty years, its energy consumption doubled in the past twenty years; and whereas the world population doubled in the past thirty years, the world energy consumption doubled in the past decade.

With our resources rapidly disappearing, a significant challenge faces science in the future. Entomologists have made and will continue to make major contributions to our understanding of the principles of biology and knowledge as applied to the needs of man. The study of the biology of insects contributes to a more complete understanding of life systems in general. Insect systematics, in particular, has made major contributions to our understanding of the taxonomy of insects, as well as other life systems, because of the talents of insect systematists and because there are more kinds of insects than all other plants and animals combined. The evidence suggests that there are between one and five million insect species.

The study of the renowned fruit fly *Drosophila* continues to be the prime model for investigations of heredity and evolution. Theories and principles of animal behavior have been enriched by studies of insect behavior. Both the diversity of insect material available and the relative ease of handling and manipulating insects contribute to their usefulness in behavioral studies. Research on the societies of bees, ants, and termites has provided a wealth of scientific information concerning the structure and organization of animal communities. Ecology, too, has benefited from entomological research. In fact, most of the principles of population dynamics have evolved from studies of insect populations (Andrewartha and Birch 1954).

The study of insects is important not only because the principles of insect systems can be applied to other life systems; insects themselves play a vital role in maintaining a quality environment for the survival of the human society. Few appreciate the significance of insects in the dynamics of our life system.

Insects are probably the most abundant animal type. Insects may exist at weights up to 2000 pounds per acre in favorable habitats (Wolcott 1937). Compare this with birds, estimated at only about one pound per acre. More importantly, compare the density of insects and humans. I estimate that insects exist at an average of about 400 pounds per acre in the United States—about 30 times the weight of mankind in the United States, where there are about 14 pounds of human protoplasm per acre.

Insects are the prime source of food for many species of fish, such as trout and salmon; amphibians, such as toads and frogs; birds, such as the bluebird and the house wren; and mammals, such as moles, shrews, and skunks. Further, insects help concentrate and convert plant and animal material into food for other animals.

Insects not only keep the life system functioning by eating and being eaten, but also serve man as plant pollinators. Despite man's many technological advances, effective substitutes for insect pollination of plants, both cultivated and in the wild, have not been found. The adequate production of fruits, vegetables, and forages depends upon the activity of bees and other insects which pollinate their blossoms. A single honeybee may visit and pollinate 1,000 blossoms in a single day. In New York State, which has about 3 million domestic honeybee colonies, each with about 10,000 worker bees, honeybees could visit 30 trillion blossoms in a day. Wild bees pollinate a number at least equal to that. Thus on a bright, sunny day 60 trillion blossoms may be pollinated by the bees in New York, a task impossible for man to accomplish today.

Several groups of insects feed on dead and decaying animals and plants and on animal wastes, thereby helping to degrade these environmental wastes. Through this type of recycling, many elements vital to man's existence are returned to the environment for reuse and the accumulation of waste on the surface of the land is avoided. This is no small accomplishment; in the United States about 1.7 billion tons of animal wastes are produced each year, about 1,500 pounds of animal wastes per acre.

In Australia the importance of insects in disposing of animal wastes was exemplified. With the introduction of cattle into Australia, where there were no dung beetles to bury the wastes beneath the soil surface, the cow pats tended to accumulate on the pastures (Waterhouse 1974). With a cow producing about 20,000 pounds of dung per year, the pastures were slowly being buried in cow pats and were unsuitable for cattle grazing. Dung beetles were introduced from Central America, Asia, Africa, and Europe, and several of the introduced species became established. Ecological balance was restored. Waterhouse called the result the "biological control" of cow dung in Australia.

Beneficial insects also control many pests of man. The biological control of pests by insects protects both U.S. and world crops. However, not all insect activity is beneficial to man—about 1% of the insects are pests. According to the best estimates available, losses in agriculture in the United States due to insects, pathogens, and weeds amount to 33% of our crops annually (U.S. Department of Agriculture 1965). Insects destroy about 13% of the crops in spite of all the insecticides used and the biological controls employed.

The usage of DDT and other synthetic pesticides has grown in the decades following their introduction in 1946 to over 1.1 billion pounds annually (fig. 3). Although overall crop

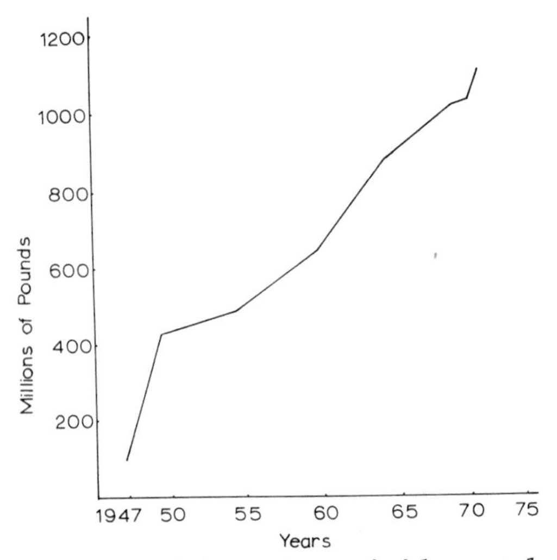

Fig. 3. Quantities of pesticides produced in the United States (data from USDA 1971, 1973).

losses due to insects have increased, important advances have been made in reducing insect losses from certain pests in some crops. For example, losses in yield and quality from potato insects declined from 22% in 1910-35 (Hyslop 1938), to 16% in 1942-51 (USDA 1954), and to 14% in 1951-60 (USDA 1965).

In contrast, overall losses in apples caused primarily by codling moth and apple maggot have not declined with increased use of organic insecticides. A 10.4% loss in yield was reported for the period 1910-35 (Hyslop 1938), a 12.4% loss for 1942-51 (USDA 1954), and a 13% loss for the period

1951-60 (USDA 1965). This loss pattern probably reflects both the higher quality standards for salable fruit and the decline in sanitation and other cultural controls formerly practiced in orchards.

According to USDA estimates, corn losses due to insects have also been increasing. A 3.5% loss was reported for the period 1942-51 (USDA 1954), and a 12% loss for the period 1951-60 (USDA 1965). Factors contributing to these losses include the culture of corn on the same land year after year, thus increasing rootworm susceptibility (Tate and Bare 1946; Hill, Hixon, and Muma 1948; Ortman and Fitzgerald 1964; Robinson 1966), and the planting of insect-susceptible corn types rather than resistant types, thus increasing corn-borer susceptibility (Painter 1951; Sparks et al. 1967; Starks and McMillian 1967). The latter factor has been implicated in the greater pest losses in rice and wheat varieties used in "the green revolution" (Pradhan 1971).

Pesticide use in agriculture, as mentioned above, has been increasing rapidly in the United States (fig. 3). Of the half-billion pounds of pesticide applied to crop and farm-lands, 52% are insecticide, 34% herbicide, and 14% fungicide (USDA 1973). According to the latest data published by the U.S. Department of Agriculture (1968 and 1972), cropland (including pastures) in 1966 totaled 890.8 million acres, of which only 5% were treated with insecticides, 15% with herbicides, and 0.5% with fungicides. If cropland devoted to pastures is removed from the total acreage, the cropland treated with insecticides, herbicides, and fungicides is about 12%, 29%, and 1%, respectively.

No published estimate is available on the extent of bioenvironmental controls employed in U.S. agriculture (PSAC 1965). However, data assembled in an informal study (Pimentel, unpublished) indicate that about 13% of the crop acreage (including pastures) in the United States is under bioenvironmental control for insects. This suggests that, although pesticides receive the publicity (both good and bad), bioenvironmental controls for insects are being practiced on nearly three times as many crop acres.

Bioenvironmental controls here refer to some action initiated by man to control pests associated with crops. However, we should not neglect the role of controls by natural enemies in limiting pest-population outbreaks. Seldom is the important role of natural enemies appreciated until these enemies are inadvertently destroyed and severe pest outbreaks seriously damage crops (DeBach 1964; Pimentel 1971; Van den Bosch and Messenger 1973).

Another significant loss in crops occurs in storage after harvest. Losses of stored foods due to insects, rodents, and microorganisms are estimated to be about 9% in the United States (USDA 1965). Nearly half of this loss is probably due to insect attack. This 4%—that is, 3.5% of the original crop—plus the 13% loss during preharvest, amounts to a total annual loss of food in the United States due to insects of about 16.5%. Total loss of pre- and postharvest crops to all pests totals about 39%!

The above discussion provides a broad perspective of the interrelationship of insects, science, and society. In a brief symposium it would be difficult to cover all aspects of this interrelationship; hence certain topics were selected to highlight the significant contributions entomology is making to science and the society of man.

REFERENCES

Andrewartha, H.G., and L.C. Birch. 1954. The distribution and abundance of animals. Univ. of Chicago Press, Chicago.

Borgstrom, G. 1973. Harvesting the earth. Abelard-Schuman, New York.

Brown, L.R. 1965. Increasing world food output—problems and prospects. U.S. Dep. Agr. For. Agr. Econ. Rep. no. 25.

DeBach, P.H. [ed.]. 1964. Biological control of insect pests and weeds. Reinhold, New York.

Food and Agriculture Organization. 1970. The provisional world plan for agricultural development. 2 vol. United Nations, Rome.

Freedman, R., and B. Berelson. 1974. The human population. Sci. Amer. 231(3):30-39.

Hill, R.E., E. Hixon, and M.H. Muma. 1948. Corn rootworm control tests with benzene hexachloride, DDT, nitrogen fertilizers, and crop rotations. J. Econ. Entomol. 41:392-401.

Hubbert, M.K. 1962. Energy resources: a report to the Committee on Natural Resources. National Academy of Sciences—National Research Council Publ. 1000-D. Washington.

Hyslop, J.A. 1938. Losses occasioned by insects, mites, and ticks in the United States. U.S. Dep. Agr., E-444.

National Academy of Sciences. 1971. Rapid population growth, vol. 1 and 2. Johns Hopkins Press, Baltimore.

Ortman, E.E., and P.J. Fitzgerald. 1964. Developments in corn

rootworm research. Proc. Annu. Hybrid Corn Industry Res. Conf. 19:38-45.
Painter, R.H. 1951. Insect resistance in crop plants. Macmillan, New York.
Pimentel, D. 1971. Ecological effects of pesticides on non-target species. U.S. Govt. Print. Office, Washington.
———. Bioenvironmental control of pests: a research assessment. Unpublished ms., prepared for the National Science Foundation.
Political and Economic Planning. 1955. World population and resources. A report by PEP. Chiswick Press, London.
Pradhan, S. 1971. Revolution in pest control. World Sci. News 8(3):41-47.
President's Science Advisory Committee. 1965. Restoring the quality of our environment. Report of the Environmental Pollution Panel. The White House, Washington.
———. 1967. The world food problem. Report of the Panel on the World Food Supply, The White House, vol. 1, 2, and 3. U.S. Govt. Print. Office, Washington.
Robinson, R.E. 1966. Sunflower-soybean and grain sorghum-corn rotations versus monoculture. Agron. J. 58:475-77.
Sen, B.R. 1965. Food population and development. Food and Agriculture Organization, United Nations.
Sparks, A.N., H.C. Chiang, C.A. Triplehorn, W.D. Guthrie, and T.A. Brindley. 1967. Some factors influencing populations of the European corn borer, *Ostrinia nubilalis* (Hübner), in the North Central States: resistance of corn, time of planting, and weather conditions. II. 1958-1962. Iowa State Univ. Agr. and Home Econ. Exp. Sta. Res. Bull. 559: 63-103.
Spengler, J.J. 1968. World hunger: past, present, prospective "That there should be great famine." World Rev. Nutr. Diet. 9:1-31.
Starks, K.J., and W.W. McMillian. 1967. Resistance in corn to the corn earworm. J. Econ. Entomol. 60:920-23.
Tate, H.D., and O.S. Bare. 1946. Corn rootworms. Nebr. Agr. Exp. Sta. Bull. 381.
United Nations. 1957-1971. Statistical yearbooks. Statistical Office, Department of Economic and Social Affairs. New York.
United States Department of Agriculture. 1954. Losses in agriculture. Agr. Res. Ser. 20-1. U.S. Govt. Print. Office, Washington.
———. 1965. Losses in agriculture. Agr. Handbook no. 291, Agr. Res. Ser. U.S. Govt. Print. Office, Washington.
———. 1968. Extent of farm pesticide use on crops in 1966.

Agr. Econ. Rep. no. 147, Econ. Res. Ser. U.S. Govt. Print. Office, Washington.
———. 1971. The pesticide review 1970. Agr. Stabil. Conserv. Ser. U.S. Govt. Print. Office, Washington.
———. 1972. Extent and cost of weed control with herbicides and an evaluation of important weeds, 1968. Econ. Res. Ser. U.S. Govt. Print. Office, Washington.
———. 1973. The pesticide review 1972. Agr. Stabil. Conserv. Ser. U.S. Govt. Print. Office, Washington.
United States Department of Labor. 1973. Consumer price index—U.S. average—general summary and groups, subgroups, and selected items. Mon. Labor Rev. 96(12):106-11.
Van den Bosch, R., and P.S. Messenger. 1973. Biological control. Intext Educational Publ., New York.
Waterhouse, D.F. 1974. The biological control of dung. Sci. Amer. 230(4):101-9.
Wolcott, G.N. 1937. An animal census of two pastures and a meadow in northern New York. Ecol. Monogr. 7:1-90.
Woodham, A.A. 1971. The world protein shortage: prevention and cure. World Rev. Nutr. Diet. 13:1-42.

PART I

Insects and Their Social Implications

INSECTS AND HUMAN WELFARE

John J. McKelvey, Jr.

INTRODUCTION

Insects nurture and protect us, sicken us, kill us. They bring us both joy and sorrow. They drive us from fear to hate, then to tolerance. At times they bring us up short to a realization of the way the world really is, and what we have to do to improve it. Their importance to human welfare transcends the grand battles we fight against them to manage them for our own ends. Most of us hate them, but some of us love them. Indeed, at times they even inspire us.

THE SILENT MAJORITY

That great silent majority of insect species around us operates for the most part in the interests of human welfare. Some figure as vital, basic natural resources for people to make use of and to enjoy. Women in Thailand, Korea, Japan, Italy, India, and China cultivate silkworms to help eke out a living and for the luxurious fabrics their cocoons furnish. Children in Africa relish termites and grasshoppers, which they catch on the wing for their sweet taste. Men all over the world keep bees for honey and wax, products essential for industry as well as for food and health. The bees also pollinate their crops. People marvel at the incredible physical feats insects perform. Their bizarre life histories, their courtship, and their sexual escapades rivet our attention upon them. Entire societies—the Egyptians with their sacred scarabs, for example—have embraced insects in their religious and superstitious beliefs.

THE UNRULY FEW

The same insects that benefit us, their relatives, and other insects as well, can and do exert a powerful negative impact on human society. Some relatives of the silkworm figure among our most devastating plant pests. Termites may bring down one's house or shred one's favorite book. Even the honeybee can be deadly: witness the havoc the African bee wrought when it arrived in Brazil (National Academy of Sciences 1972).

Our task is to recognize when an insect becomes a pest, to banish those insects which, as pests, are doing us more harm than good and which we therefore feel we can do without, and to learn to live with those which, though doing some harm, also are of benefit to us.

Managing the unruly few becomes ever more challenging and expensive as we try to combat disease and malnutrition, increase food and fiber production, and enhance our social, cultural, and economic life, and while we let our own population rise from 3.5 to 8 billion people, inexorably approaching the predicted maximum carrying capacity of this planet in terms of land, living space, and quality of life. This maximum will be reached shortly after the year 2000 (University of California Food Task Force 1974).

Of the pests, we shall first consider the tsetse flies, the black flies, and the mosquitoes—vectors of diseases which have special impact on development in both the humid and the dry tropics, areas which cover much of the developing world. The diseases are: sleeping sickness in man and nagana in cattle, both of which tsetse flies transmit; river blindness, prevalent in Africa and Central America; and malaria, virtually worldwide in distribution. Next we shall move to the problems of protecting crops and livestock in developing countries to achieve maximum production; and then we shall turn to a prime topic in industrialized nations: bringing our chemical technology in insect control into phase with human welfare.

HEALTH

More than ever before, the human race needs the humid and the dry tropics for living space and for producing foods and fibers, yet the flies and mosquitoes that spread human and animal diseases still prevent use of potentially productive areas of these regions. In the humid tropics of Africa, for example, the tsetse fly, because it transmits sleeping sickness to people and nagana to cattle, still dominates 4.5 million square miles. While the control for sleeping sickness may be adequate momentarily, the tsetse flies all but prevent the people who live in the humid tropics from raising cattle for food and from using oxen as beasts of burden.

Tsetse flies also frequent the river courses of the dry tropics of Africa, where the people and their cattle are targets for tsetse bites. Man-fly and cattle-fly contact is particularly intense at the watering places. During the dry season the women congregate at water holes in the riverbeds, waiting in queues to fill their water jugs, and exposing themselves to hungry tsetse flies, which feed on them and inoculate them with deadly trypanosomes; the men likewise herd their cattle there to drink (Nash 1944). By killing people and decimating livestock, tsetse protects its own ecological niche: the fly, with its trypanosomes, lightens the population load in the area, in this way slowing down man's destruction of its habitat, "the bush," for agricultural purposes.

For a hundred years explorers, missionaries, entrepreneurs, medical men, and entomologists have subjected this insect to the full range of control measures devised to combat insects. The technology is probably available today to wipe tsetse out, but for a variety of reasons implementation of that technology fails to get the job done. Nagana is rife in cattle. Today sleeping sickness in man lurks in the shadows, killing a few thousand persons each year; tomorrow it may well break out again in epidemic proportions, as it has periodically done in the past.

The female black fly, when she bites human beings, transmits the filarial worms that cause river blindness, onchocerciasis. A major cause of incapacity in the savanna areas of West Africa, this insect prevents human settlement and agricultural development of many fertile river valleys where it lives and breeds. In 1974 seven countries—Dahomey, Ghana, Ivory Coast, Mali, Niger, Togo, and Upper Volta—in cooperation with the United Nations Development Programme, the Food and Agriculture Organization, the International Bank for Reconstruction and Development, and the World Health Organization (WHO), agreed on a strategy for controlling the fly and the disease in the Volta River basin and for promoting subsequent economic and agricultural development of the area. The project area, which encompasses approximately 440 thousand square miles, is inhabited by almost 10 million people, one million of whom are infected with river blindness. Of these, 70 thousand are considered, for economic purposes, to be blind. The program—planned for a twenty-year period—will involve spraying the biodegradable insecticide Abate, from the air onto the water, to reach the black fly's breeding sites in the fast-flowing streams of the river basin.

The anopheline mosquitoes, in transmitting the causal

agents of malaria, maintain that disease as a major health problem for the growing populations in the tropics and subtropics. In 1973 malaria was endemic to 145 of the 209 countries or territories for which data are available; Africa south of the Sahara is the most important locus of this disease. There 196 million persons are subject to malarial risk. In any one year there may be 100 million cases. No satisfactory vaccine is available; implementation of mass chemotherapy is not completely effective, nor is it practical. As a consequence, people will have to rely on the use of insecticides to kill the mosquito vectors well into the foreseeable future.

DDT remains the insecticide of choice for combating malarial mosquitoes. Forty thousand tons of technical DDT per year will be required for this purpose over at least the next decade. Resistance of mosquitoes to DDT has emerged in approximately one percent of the areas under treatment. During the next twenty years, according to Dr. J.W. Wright of WHO, research will continue to be essential, directed toward (1) making the most effective use of insecticides presently available, (2) developing new and safe chemicals to control resistant strains of mosquitoes, and (3) designing imaginative plans to integrate all available measures into one overall operating strategy (Wright, in press).

The discovery of DDT, with its high order of effectiveness and residual qualities when applied to the internal wall surfaces of houses, brought optimism, shortly after World War II, that the mosquitoes that transmit malaria could be not only controlled, but eradicated. Today, it seems clear, *Man's Mastery of Malaria* (Russell 1955) has yet to be achieved.

Sleeping sickness, river blindness, and malaria, as vector-borne diseases of man, and nagana, as a vector-borne disease of cattle, constitute but four of the diseases that stand as guardians of the tropics against man. One would have to be unusually callous to argue that the flies and mosquitoes that transmit these diseases should be permitted to "do their thing." Yet people do make allowances for these insects as pests. Some environmentalists and wildlife enthusiasts applaud the existence of tsetse, pointing out that, were it not for these flies, which have deterred man from keeping cattle in extensive areas occupied by wildlife, the great game resources of the African continent would have disappeared long ago. One might also argue that in their long association with the malarial parasite the people in Africa have gained a measure of resistance to malaria and that, therefore, the use of pesticides, always with unfortunate side effects in an environment, may not be urgently required for the control of the mosquito vectors of the disease.

People can, however, exaggerate the potentially damaging effect of pesticides used to combat flies or mosquitoes, especially in areas in which they themselves do not live and may never have suffered. They may encourage the foreclosure of use of an effective pesticide before a substitute is available which is harmless to species other than the target pest. They might do well to reflect that the people in daily contact with these pests would prefer not to contract a case of sleeping sickness from tsetse, or to mix a case of malaria from mosquitoes with the business or the romance of canoeing down the Congo River in a dugout in Zaire.

Dr. J.W. Wright (in press) of WHO points out:

> Governments must begin long-term programs based upon cadres of well-trained scientists and field workers, on a sound knowledge of the interrelationship between man, the vector, the organism and the reservoir, where one exists, and the judicious use of pesticides so that the highest level of control can be achieved compatible with minimum contamination of the environment and the maximum safety to man, animals and other non-target organisms.

A vast amount of research has been performed, with brilliant results, on insect vectors of the diseases of the tropics in years past, but clearly this research in no way mitigates the need to accelerate and to intensify work on these pests in the future.

AGRICULTURE

In spite of vector-borne diseases that plague him, man increases his numbers, expands into new territory in the developing world, and disseminates those unshakable pests that tag along on his crops and livestock. These insects find themselves in new and different milieux; there they interact with local insects, plants, and animals and create new relationships which entail new and different problems. These problems must be solved in the context of a world energy shortage which affects the cost, manufacture, and even availability of insecticides. They place an especially heavy onus on the economic entomologist, particularly in developing countries, to seek ways of protecting crops without the use of insecticides. Breeding plants resistant to pests holds promise, but even this method has its drawbacks. High-yielding, resistant varieties may not have nutritional value equivalent to that of standard varieties; and farmers are not likely to accept and to plant varieties of high nutritional value—high-lysine corn, for example—until insect resistance and good agronomic characteristics are bred into them.

The developing countries need strong research and training bases for entomologists at both international and national levels. A network of international research centers has been created in the Philippines, Mexico, Colombia, Peru, Nigeria, India, Ethiopia, Kenya, and Taiwan during the last decade in response to the need and opportunities for increasing plant and animal production (Consultative Group on International Agricultural Research 1974). Each research center has its special program; collectively they cover work on rice, wheat, corn, sorghum, millet, grain legumes, roots and tubers, farming systems, and animal health and husbandry. These centers employ entomologists to lead research, and they provide training opportunities for aspiring young professionals.

Crop- and animal-oriented institutes, however, can undertake only part of the investigations required for a full understanding of insects. Basic research must back up applied research. One institute which will address this need is the International Centre for Insect Physiology and Ecology (ICIPE) recently established in Nairobi, Kenya. Its program is organized around five principal pests affecting man in East Africa: termites, armyworms, mosquitoes, tsetse flies, and ticks. Its administrative structure assures that its work will be of international use, yet relevant to African problems. Distinguished scientists who work at ICIPE may, for example, discover a wide pool of tick genes that would be useful in their efforts to manage this arthropod population; or they may find out why certain trypanosomes have to pass through the salivary glands of tsetse flies before they can become infective to cattle. These and other basic investigations may lead to new and imaginative concepts for managing pest populations (ICIPE 1973).

While the developing nations strive to bring their agricultural-production capabilities into line with annual population growth rates as high as 3.4% (University of California Food Task Force 1974), people of the so-called developed countries voice concern about environmental pollution associated with the use of chemicals to achieve maximum production. Agricultural analysts expect that the United States, with its population probably stabilizing at 300 million by the year 2000, will continue to produce grains and fibers in surplus and thereby continue over the next quarter of a century in its traditional role as an exporter of agricultural products (University of California Food Task Force 1974). It remains to be seen how our technology of insect management can be made to accommodate the public interest to maintain health and environmental quality and still effectively control the insect pests of our crops and livestock.

For 150 years in the United States, we have been blessed or cursed with a "culture of cotton" which today, because of the persistence of the insect pests that feed upon the plant, accounts for more than half of the insecticide load applied annually to all crops in this country. Some people would argue that a simple agricultural-policy decision to discourage cotton production would be in the best interests of the country. Others take the view that cotton production is essential to our economy.

Long before persistent nonbiodegradable pesticides became an issue, however, one insect, the boll weevil, wrought marvelous institutional changes in agriculture and caused conceptual advances in agricultural services. Texas, in the first decade of the twentieth century, was the battleground for the fight against the boll weevil, which had crossed the Mexican border and was devastating cotton fields. There a USDA representative, Dr. Seaman A. Knapp, established a demonstration farm where he showed farmers how to produce satisfactory cotton crops despite the boll weevil. He recommended careful seed selection, deep plowing in the autumn, wide spacing of plants, intensive cultivation, systematic fertilization, and rotation of crops.

One of the early Rockefeller boards, the General Education Board, in search of a method of improving the lot of the southern farmer, joined with Dr. Knapp and the USDA in a program that began in a few rural centers in six southern states in 1906 and multiplied to nearly 100 thousand demonstration farms throughout the South. By 1914 the program expanded to include corn and other crops. These model farms helped pave the way for the nationwide agricultural-extension services established a few years later (Fosdick 1952).

In similar fashion, institutional adaptations are taking place today because of the need for new and refined methods of plant protection. At Cornell University leaders in entomology, plant pathology, and plant breeding are initiating a program that marshals their combined talent, experience, and resources to enable corn to defend itself against its many pests and pathogens (Brody 1975). Entomologists at universities and experiment stations in New York, Texas,

Mississippi, Illinois, and California are rapidly strengthening similar cross-departmental, multiuniversity ties.

The boll weevil—still with us—together with the bollworm and the pink bollworm, continue to challenge the entomologist to find new ways to cope with their depredations. Tactics based on population dynamics, theory and practice, which include the technique of liberating sterile male weevils to compete with fertile males in nature, undergird a pilot scheme to oust the boll weevil from the United States. In spite of the battery of methods being applied in integrated fashion to bring the weevil's population below the critical mass necessary for it to sustain itself as a species, only partial success seems likely.

As for the bollworm, a general feeder which is just as serious a pest of corn as it is of cotton, breeding cotton resistant to its attack may turn up new cottons which will give a "quantum jump" in yield and quality, as well as protection from insect attack. After all, the attempt to breed cotton resistant to another enemy, the fleahopper, resulted in experimental lines that combine resistance to fleahoppers, early maturity, and the ability to escape bollworm attack, without the benefit of insecticides, to outproduce commercial varieties by nearly 100%.[1]

With the pooling of disciplinary resources has come a wider and more flexible outlook. In recent years researchers have come to recognize the importance of a plant-protection method that may give 10 to 20% yield increases, whereas earlier targets might have been 50 to 90%. The rigid concept of pest control, total kill, is not consonant with modern knowledge, with biological possibilities, or with human welfare. We cannot yet "get" that last tsetse fly, for example, no matter how hard we try.

The burden of my message, then, is that world progress in improving our food, health, and quality of life will dictate that we constantly redefine which insects we consider to be pests, and for any one insect, when that insect is a pest and when it is not. This means: (1) evaluating on a sliding scale, so to speak, long-term versus short-term impact of insects on human activities, (2) managing the control measures at our disposal and any new ones we can devise within the limits of human acceptability and environmental protection, and (3) reexamining our views on the extent of damage we are willing to permit insects to cause to our health and to our crops before we take action.

---

[1] P.L. Adkisson 1974: personal communication.

THE PUBLIC AND THE PROFESSION

The layman possesses a tacit sensitivity to, and appreciation for, insects as they enhance cultural, social, and economic life. At times, persons in all walks of life express these responses spontaneously. Truck drivers, their vehicles backed up behind a landslide on a road on the eastern slopes of the Andes in Peru, whiled away the time helping me catch insects. A machinist has welded together a facsimile of a butterfly, using washers for the wings, a cut steel flooring nail for the body, and finishing nails for the antennae. A stamp collector seeking ways of bringing organization into a chaotic collection may decide to specialize only in those stamps which bear the image of insects, imprints from countries proud of their insect heritage.

Authors develop plots for their stories around insects; *The Gold Bug* assured Edgar Allan Poe a place in literature as the father of the modern detective story. And the audacity of a cockroach in plain sight happily consuming the paste that binds together a collage in a work of art forty-seven stories above ground level in a modern steel, stone, and glass skyscraper in New York City ought to inspire some would-be poet to outdo Robert Burns, who philosophized in his poem "To a Louse" on the impudence of that insect gracing the hat of a woman in church and was inspired to conclude:

> O wad some Power the giftie gie us
> To see oursels as ithers see us!
> It wad frae mony a blunder free us
> An' foolish notion:
> What airs in dress an' gait wad lea'e us,
> An' ev'n devotion!

Even soldiers on the battlefield have their moments for appreciating the beauty of insects. The classic story (Howard 1930) I like to quote about Count Pierre François Aimé Auguste Dejean, an aide-de-camp to Napoleon I, bears this out. He "never lost a chance to add an insect to his collection. . . . At the battle of Alcanizas in Spain, which Dejean won after a very hard fight in which he took a great number of prisoners, he suddenly saw near a little brook a brilliant and rare" beetle resting on a flower.

> At the head of his troops, facing the enemy, Count Dejean was about to give the signal to charge, but, seeing the insect, he at once dismounted, captured it, pinned it in

> his helmet, remounted his horse, and gave the order for one of the most vigorous charges of the campaign. After the battle he found that his helmet had been "horriblement maltraite" by a cannon-ball, but he recovered his precious beetle intact.
>
> All of the soldiers in his regiment learned to collect insects. Each carried a small vial of alcohol in which to place the insects he collected. Even the enemy knew of Dejean's "eccentricity." . . . "Those who found dead soldiers on the field having with them a little bottle containing insects in alcohol" always carried the bottle to Dejean, regardless of who won the battle.

The profession of entomology needs an interested and a participating public; entomologists must maintain and strengthen communication with that eager constituency which they serve even though a layman's logic can sometimes be extremely exasperating. One farmer I knew in the Hudson River Valley became overly impressed with the reproductive capacity of the European corn borer which was severely damaging his sweet corn. His homespun approach to cutting down the insect's population showed more spirit than sense. Because it cost so much, he thought he could not afford to dust his corn with insecticide to kill the borers. But, he said, "When I drive a truckload of my corn to the Albany market, if I see a caterpillar crossing the road ahead I swerve to smash it. It might be a corn borer."

Finally, the elm spanworm, a common inchworm, the kind that a child will hold and admire on a twig, tells us much. (This insect takes its name from the American elm, hardly a source of inchworm food and shelter anymore. Another pest, a beetle, teamed up with a fungus and removed that tree from the American scene. With it is fading into history the image of small-town America with Main Street under the shade of elms.) Normally, the spanworm, feeding on a variety of trees, lives in obscurity, but periodically its numbers get grossly out of control. So it happened in the late 1960s and early 1970s. As the spanworm procreated, it ate its way to notoriety in the suburbs of the eastern seaboard.

Few of the gardens in the area planted for food suffered damage, but in early June the insect had completely defoliated the maple, cherry, and oak trees. To keep a well-manicured lawn meant fighting one's way through countless threads of silk. You could hear the inchworms chewing, and a backyard picnic was out of the question because of the streamers of silk and the droppings of the larvae.

The concern of the suburban householder for preservation of birds and wildlife is great; insecticides, noxious and persistent, are frowned upon, and many persons refrain from using them. But these same persons yielded in the face of the exasperation and the inconvenience the spanworm caused them and applied such insecticides, even at a cost of $25 to $50 per tree. One year, two years passed. Still the sprays did not constrain this pest. Like snowflakes in July, filtering through the leafless trees in the twilight, the moths appeared each summer to lay their eggs for another go at the same trees the following year.

Eventually, *Ooencyrtus* sp., an insect too small and too insignificant to be noticed by anyone but an entomologist looking for it, set to work and, as it had done in earlier years, parasitized the eggs of the spanworm. Where impatient men with modern methods failed, it brought the spanworm population back to normal (Kaya 1974).

Insects? We hate them! Though at times some of us "love" them. It's true that their importance to human welfare transcends the grand battles we fight to manage them for our own ends. Sometimes we stand in awe of insects for what they do normally in coping with the pests that offend us, when they reveal their normal role, as did the tiny egg parasite which several years in a row happened to go off duty and let a lowly inchworm get out of hand. And sometimes we undertake research to find answers to the questions they pose. Indeed, insects do inspire us.

REFERENCES

Brody, J.E. 1975. A corn resistant to two major pests is discovered. New York Times, Jan. 3:20.

Consultative Group on International Agricultural Research. 1974. International research in agriculture. New York.

Fosdick, R.B. 1952. The story of the Rockefeller Foundation. Harper & Brothers, New York.

Howard. L.O. 1930. A history of applied entomology (somewhat anecdotal). Smithsonian Miscellaneous Collections, vol. 84. Smithsonian Institution, Washington.

International Centre of Insect Physiology and Ecology. 1973. First annual report, 1973. Nairobi.

Kaya, H.K., and J.F. Anderson. 1974. Collapse of the elm spanworm outbreak in Connecticut: role of *Ooencyrtus* sp. Environ. Entomol. 3:659-63.

Nash, T.A.M. 1944. A low density of tsetse flies associated with a high incidence of sleeping sickness. Bull. Entomol. Res. 35:51.

National Academy of Sciences, Committee on the African Honeybee. 1972. Final report. Washington.

Russell, P.F. Man's mastery of malaria. 1955. Oxford University Press, London.

University of California Food Task Force. 1974. A hungry world: the challenge to agriculture. Summary report. College of Agriculture, Division of Agricultural Sciences.

Wright, J.W. Insecticides in human health. Insecticides for the future: needs and prospects. Proceedings of a Rockefeller Foundation Conference, Bellagio, Italy, April 22-27, 1974. In press.

THE EVOLUTIONARY SIGNIFICANCE OF THE SOCIAL INSECTS

Edward O. Wilson

The deeper biological significance of social insects lies in their status as a major, independent experiment in social evolution. We may treat them virtually as newly discovered social beings on some alien planet. The termites, ants, and most social bees and wasps, which constitute the eusocial forms by virtue of the possession of sterile castes and cooperation between overlapping generations, are representatives of the arthropod superphylum, which stands far apart from our own phylogenetic group, the "echinoderm" superphylum.

The ancestors of the two groups diverged in evolution more than 600 million years ago. The progenitors of the insects and tetrapod vertebrates then invaded the land separately, accommodated to the new, harsh environment each by its own idiosyncratic physiological techniques, diverged to the opposite ends of the size spectrum, and came to occupy largely different ecological roles. The final result is that an ant and a man are radically different in embryology, life cycle, physiology, and organization of the nervous system. Any superior extraterrestrial observer would be likely to comment on that fact before turning to the subject of man's modest intelligence.

By scanning all social groups from colonial bacteria and protozoans to the social mammals, our observer would be bound to place the social insects about in the middle of the scale of social achievement. If one judges sociality by the usual intuitive qualities of altruism, cooperativeness, division of labor, and group cohesiveness, the social insects fall well below corals and other colonial invertebrates. But they range high above the social nonhuman mammals, including the monkeys and apes. This ordering creates something of a paradox, because it is the reverse of the phylogenetic order. As a rough rule, the more primitive and older the group, the more advanced are its social species. Also, the acquisition of intelligence is no guarantee of the evolution of social behavior.

The key to this inverse relationship appears to be kinship. The colonial invertebrates, due to their primitive body plan, are able to reproduce rather easily by budding or fission. The result is a clone of genetically identical individuals. To sacrifice oneself for a clonemate by turning into a sterile zooid may be no sacrifice at all in the genetic sense, because the clonemate is a genetically identical replicate. All that is needed to spread genes favoring the altruistic behavior through the population of clones is that at least one member of the colony leave more than twice the number of offspring than would be the case if there were no altruism. This basic view of social evolution by kin selection was first developed by W.D. Hamilton (1964). In Hamilton's terminology, the inclusive fitness of the clonemates—that is, the offspring left in the next generation by both the altruist and the beneficiary—exceeds the individual fitness of the same number of zooids who do not engage in the altruistic relationship.

By the same token, vertebrates are not expected to be very self-sacrificing, at least not to groupmates other than their own offspring. Since they reproduce almost exclusively by conventional sexual means, with heterogametic or heterogenic sex determination based on a diploid inheritance system, the largest fraction of genes that two individuals can share by common descent is one-half. This is not always strictly true, since monozygotic siblings share all of their genes. But it is close enough to explain why true altruism beyond parental care is a rarity in vertebrates. Vertebrate societies are characteristically strife ridden, and sterile castes are unknown.

Human beings have nevertheless reversed the overall phylogenetic trend toward lower sociality. They have done so not by reducing selfishness, but rather by the employment of intelligence in devising alliances and compromises with their competitors. Aided by a powerful, syntactical language and a keen intuitive understanding of kinship—the ties of "blood" by which some degree of unfairness in others is tolerated—human beings establish contracts that may extend over generations. This arrangement permits a division of labor within societies exceeding that of the most complex invertebrate colonies. Nevertheless, the division is based on truly vertebrate roles, which normally contribute more to individual genetic fitness than to group fitness. It is apparently never founded on altruistic castes.

The social Hymenoptera, comprising the ants, bees, and wasps, are almost exactly intermediate between the colonial invertebrates and the vertebrates in the degrees of kinship possessed by groupmates. Hamilton was the first to point out that the haplodiploid mode of sex determination in the Hymenoptera (males from unfertilized, haploid eggs; females from fertilized, diploid eggs) results in females being more closely related to their sisters than to their own offspring. More precisely, females share by common descent 75% of their genes with their sisters but only 50% with their daughters.

This result has been used to explain the statistical dominance of advanced sociality by the Hymenoptera. Although less than 20% of all insect species have belonged to this order throughout the Cenozoic era, eusociality has arisen at least eleven times within the Hymenoptera: twice or more in the wasps (Stenogastrinae, Vespinae, and probably in the sphecid genus *Microstigmus*), eight or more times in the bees, and once or perhaps twice in the ants. Throughout the rest of the insects, eusociality is known to have originated in only one other living group, the termites (Wilson 1971).

The central argument is that the bias due to haplodiploidy predisposes the evolution of societies in which sisters behave altruistically toward each other but in which males are uncooperative and more often rejected by their nestmates. Hymenopteran societies are, in fact, based on sisterhoods and parasitic associations of brother drones.

Recently, R.L. Trivers (in press) has refined Hamilton's conjecture in a way that permits the first rigorous testing of kinship theory. The haplodiploid hypothesis had been challenged by Michener and Brothers (1971) and by Alexander (1974) who, while not denying the logic of the basic argument, held that insect societies could evolve equally well by the exploitation of some individuals by their close relatives. Certain females, for example, might dominate their sisters to the point of subordinating them as sterile castes; or mothers could employ physiological castration or psychological control on their daughters to achieve the same results. Trivers has shown how to discriminate between the kin selection and exploitation hypotheses.

The full argument cannot be presented here, but an essential piece of it will suffice. In the case of species in which a singly inseminated queen produces all the offspring of the colony but the workers control the sex ratios, females will be related to their sisters by about three-quarters (75% of genes shared by common descent) and to their brothers by one-quarter. If exploitation by the mother queen is the prevailing factor, the colony investment in production of reproductive males and females should be 1:1, the result of Fisherian selection toward a balanced sex ratio. In other words, the summed dry weights of the sexual males and females should be about equal. If, on the other hand, kin selection is the prevailing factor, the workers should invest about three times as much in females as males, because the ratio of kinship is 3/4:1/4. In fact, measurements of many species of ants and social bees have revealed that the investment is biased toward the females at or near the expected ratio of 3:1. Solitary bees, in which of course the mother is in control, invest in the expected Fisherian ratio of 1:1. The same is true of the one slave-making ant species (*Leptothorax duloticus*) for which data are available.

Thus it appears that social insects, by being intermediate between the colonial invertebrates and the vertebrates in the degree of kinship of groupmates, have provided a unique opportunity to test and extend basic kinship theory. They have also permitted the first real test of Fisher's theory of the evolution of the sex ratio.

Another important and distinctive trait of insect societies is the stability of their caste systems. A case in point is the genus *Camponotus*, which has maintained essentially the same form of the minor and major worker castes since at least as far back as the Oligocene. An even more striking example is *Oecophylla leakeyi*, a weaver ant found in the Miocene of Kenya. The species was described from a colony fragment, the only animal society so far discovered that was preserved as a unit (Wilson and Taylor 1964). Not only are the morphological differences between the minor and major worker castes about the same as in the two living species of the genus (*O. longinoda* and *O. smaragdina*), but their numerical proportions match. Both the morphological and the statistical traits are unique among living ants.

In the course of formulating a theory of caste evolution based on linear programming models (Wilson 1968), I hit on a surprising result that seems to dictate an increasing amount of caste stability in highly social animals, such as ants and termites, that evolve by colony-level selection. The argument can be summarized as follows: The optimum caste ratio in a two-caste system is the intersection of the contingency curves, which define the various ratios of the two castes that suffice for the minimum performance of the two most important tasks on which the castes are respectively specialized. The more specialized the castes, the more nearly parallel are the contingency curves to the axes (which give the numbers of individuals in each caste).

In other words, as the number of individuals of one very

specialized caste (say, caste A) is reduced by a small amount, the number of individuals of another highly specialized caste (B) required to perform task A' increases greatly. As a result, changes in the environment may shift the intersection of the two contingency curves, hence changing the optimum ratio, but it is not likely to shift one curve a sufficient distance to completely enclose the other curve and eliminate one of the castes. In the theoretically most extreme case of specialization, the contingency curves are exactly parallel to the axes. Therefore, no matter how far the curves are shifted by modifications of the environment, they will always intersect somewhere; the optimum ratio will always contain some number of both castes above zero.

The following larger conclusion appears to be justified. The further castes diverge in the course of evolution, the more stable they become. Species that evolve to any degree by colony-level selection will tend to be locked into a particular social system. Great stability and conservatism should characterize the evolution of such systems, and this indeed seems to have been the case in the social insects.

The final feature of social insects deserving emphasis is their great diversity. There are, for example, more species of ants than all kinds of birds and mammals, both social and nonsocial, and more species of ants in a single square kilometer of Brazilian forest than all the species of primates in the world. The insects offer an imposing array of social organizations for study and comparison. The full sweep of social evolution, from the purely solitary to advanced eusocial states, is displayed independently and repeatedly by such groups as the halictid bees and sphecid and vespid wasps. In fact, there are so many species in each of the principal evolutionary grades that the entomologist can analyze them statistically, partialing out correlates in the search for the genetic and environmental determinants of social evolution. This work is in its earliest stage. The vast majority of the more than fifteen thousand species of eusocial hymenopterans and termites are unknown beyond the scientific names that label them, and thousands more undoubtedly remain to be discovered by the taxonomist. As more examples of evolutionary convergence come to light, the controlling ecological factors of social evolution can be identified with greater confidence than is possible in the vertebrates.

The climactic example of social convergence is provided by the termites and the ants. The termites arose no later than the early Cretaceous period from cockroachlike ancestors not far removed from the living wood-eating cockroaches of the genus *Cryptocercus*. The ants probably arose in early or middle Cretaceous times from tiphioid wasps not far removed from the living members of the genus *Methocha*. Thus termites and ants originated from ancestors almost as far apart as possible within the true insects. Both participated in explosive adaptive radiations between fifty and a hundred million years ago, and both have essentially stagnated at the level of the taxonomic genus and tribe during the past fifty million years. The two groups became ecologically dominant, dividing large portions of the soil and arboreal environments between them. Thus to a large extent the termites and ants represent two simultaneous experiments under largely similar ecological conditions. Only in their diet do they differ in any basic way.

The amount of parallel evolution in termites and ants is astonishing. The castes are similar in kind and form, despite the fact that workers are wholly female in the Hymenoptera and both female and male in the termites. Caste determination is under the inhibitory control of pheromones in some species in both groups. Alarm and recruitment communication, mediated principally by pheromones but also by touch and sound, are closely convergent in many details. Trophallaxis, the exchange of liquid food, is well developed in most termite and ant species, and the details of solicitation and donation are at least superficially similar. Nest construction is often elaborate, clearly adapted to enhance the regulation of internal humidity and temperature, with the most advanced termites (Macrotermitinae) holding an edge in sophistication over the most advanced ants (*Oecophylla* and perhaps some of the carton-building *Azteca*).

Notable differences also exist. The nymphs of the lower termites are major elements of the labor force, but the slug-like larvae of ants can do no more than regurgitate back some of the nutrients they receive from their adult nurses. The reproductive termite male remains with the queen and inseminates her repeatedly, whereas the male ant fertilizes the queen ant once during the nuptial flight and dies soon afterward. But these are largely details in certain mechanisms. The degree of complexity of the society, the principal means of communication, and the basic life cycle of the two kinds of social insect appear to be very close overall. A great deal more research is needed to compare the societies of termites and ants in detail, and the meaning of the parallel evolution can be profitably explored to a much greater depth.

In his essay "Possible Worlds" J.B.S. Haldane suggested that in order to gain a proper perspective of our own world we should try to conceive of alternative realities, geometries, and biological qualities that may seem fantastic at first but are nevertheless rationally conceivable. By this means the reality directly before us can be judged as a special case in a universe governed by general laws that would otherwise fail to attract our attention. Within the realm of sociobiology the insect societies have made this task much easier. In many respects their reality is stranger than any we might have imagined in the abstract. Their study will continue to shed light on the most basic processes of evolution.

REFERENCES

Alexander, R.D. 1974. The evolution of social behavior. Annu. Rev. Ecol. Syst. 5:325-83.

Haldane, J.B.S. 1927. Possible worlds. Chatto & Windus, London.

Hamilton, W.D. 1964. The genetical theory of social behavior. I and II. J. Theor. Biol. 7:1-52.

Michener, C.D., and D.J. Brothers. 1971. Were workers of eusocial Hymenoptera initially altruistic or oppressed? Proc. Nat. Acad. Sci. 68:1241-45.

Trivers, R.L. Haplodiploidy and the evolution of the social insects. Science, in press.

Wilson, E.O. 1968. The ergonomics of caste in the social insects. Amer. Natur. 102:41-66.

Wilson, E.O. 1971. The insect societies. Belknap Press of Harvard University Press, Cambridge.

Wilson, E.O., and R.W. Taylor. 1964. A fossil ant colony: new evidence of social antiquity. Psyche 71:93-103.

PART II

Communication Among Insects

NATURAL SELECTION AND SPECIALIZED CHORUSING BEHAVIOR IN ACOUSTICAL INSECTS

Richard D. Alexander

INTRODUCTION

Specialized chorusing behavior refers to rhythmical interactions, such as alternation or synchrony of successive song phrases, by neighboring calling males. Included are the massive synchronies sometimes achieved by large populations of cicadas, crickets, and katydids (Alexander and Moore 1958; Alexander 1960, 1967; Walker 1969). Some male fireflies also flash in synchrony (Buck and Buck 1968; Lloyd 1971, 1973*a*, *b*; Otte and Smiley, unpublished), and some anuran amphibians alternate (Rosen and Lemon 1974); their behavior may be considered part of the general problem analyzed here.

The choruses of acoustical insects and anurans, and the mass synchronies of fireflies, involve adult males. These males are producing signals which attract sexually responsive females. This conclusion can be drawn for all chorusing species, both from the circumstances of their performances and from comparing them with more thoroughly studied related species that do not chorus (for references see Alexander 1967; Spooner 1968; Morris, Kerr, and Gwynne, in press; Otte and Loftus-Hills, unpublished). That signals produced in chorus do attract females has been demonstrated for such species as the synchronizing snowy tree cricket, *Oecanthus fultoni* (Walker 1969), and two of the three chorusing species of seventeen-year cicadas (*Magicicada* spp.) (Alexander and Moore 1958, 1962).

The cicada signals have been shown also to attract males, as have those of the synchronizing katydid, *Orchelimum vulgare* (Morris 1971, 1972). In other chorusing species, such as the true katydid, *Pterophylla camellifolia*, there is evidence that a male's calling may cause unusually close male neighbors to move away (Shaw 1968). Whether or not more distant males in such cases tend to approach a calling male is not yet known. Attraction of both males and females to synchronous swarms has been demonstrated in fireflies (Lloyd 1973*b*).

Adult males in polygynous species are among the most competitive of all possible classes of individuals, and what they are competing for most intensively are females. That they carry out acoustical interactions, such as synchrony in chorus, giving the superficial appearance, at least, of cooperative behavior, and especially that they aggregate and chorus at breeding time, has to arouse our greatest curiosity and puzzlement.

It has not always been so. Only a few years ago most biologists referred to insect choruses as mating assemblies and supposed that a sufficient explanation had thereby been provided. There were two reasons for this attitude. First, we had not yet considered the probable outcomes of conflicts between the effects of selection at different levels in the hierarchies of organization of living matter: genes, chromosomes, genotypes, families, social groups, populations, and so on. Second, we casually and constantly regarded function at the population or species level, as we tend to see it in our own social, economic, and political affairs.

These attitudes are changing, thanks to a series of revolutionary events in biology beginning with R.A. Fisher's announcement in the 1958 revision of his 1929 book, *The Genetical Theory of Natural Selection*, that selection as he had been referring to it explains attributes only insofar as they benefit their individual possessors, and not insofar as they benefit the population or species. That should have been an explosive revelation to ecology, behavior, population genetics, and evolutionary biology in general, but for some reason it was not. Wynne-Edwards (1962) exposed the significance of Fisher's remark by generating the controversy which has since surrounded the concept of group selection, and Williams (1966*a*) developed the first serious argument against group selection as a general explanation of the traits of organisms. Two other major components in social theory were added when Hamilton (1964) and Trivers (1971) developed theories accounting for the evolution of much of what looks like cooperative or altruistic behavior—namely, kin selection and reciprocation—by natural selection acting principally at or below the individual level.

It is no longer possible casually to regard social groups of animals as functioning somehow to assist the reproduction of the population or species, and those cases like chorusing behavior in which functions at individual or genic levels have not been clearly established require our close scrutiny.

The revolution in biology which is reflected in this change of attitude has literally pushed the study of social behavior back into an almost entirely theoretical state, because most of the previous work was done with inadequate models. This means that much of this early work will have to be done over again, even if one defines social behavior so broadly as to include most of the field of behavior (see Alexander 1974; West-Eberhard, in press). In regard to the study of insect acoustical behavior, for example, the entire topic of sexual selection has failed to become a prominent aspect of our theorizing because we have not concentrated on the effectiveness of selection at the individual level.

This paper is largely a reinterpretation of my own earlier writings on chorusing behavior (Alexander 1960, 1967), in light of the changing attitude toward selection, and a review of the recent literature on chorusing.

In 1967 (p. 512) I suggested the following four possible explanations for chorusing:

> (a) Elaboration of the role of auditory feedback in rhythmical, long-continued calling of individual males may have incidentally rendered them "captive" phonoresponders to close neighbors in many species; (b) two or more males singing in alternation or synchrony may produce sound more regularly for longer periods of time (make up a more stable sound-producing unit) than lone-singing males because of stimulative and inhibitory effects upon one another . . . ; (c) phonoresponses may assist in the formation and maintenance of aggregations of males in situations in which individual males increase their chances of securing mates by joining such groups; and (d) rhythmic interaction (chorusing) by numerous males within a restricted area may prevent obscurement of the species-specific and female-attracting portions of the song or even enhance their distinctiveness.

The major shortcomings of these hypotheses are three: (1) There is no clear distinction between evolved functions and incidental effects (Williams 1966*a*); (2) there is a tendency to invoke physiological limitations to explain traits; and (3) there is no emphasis on competition between individual males, which seems to be at the heart of all male-male acoustical interactions.

The first hypothesis implies that phonoresponses, whether advantageous or disadvantageous, cannot be avoided. The second hypothesis implies that individual males cannot evolve to be stable sound-producing units, and it suggests that they assist one another, without explaining what benefits each might receive from such cooperation. The third hypothesis does not distinguish between advantages to joining males and the advantages to the already grouped males whose signals they may be using to join the group; such joining may be advantageous to one but disadvantageous to the other (see Otte's 1974 definition of *communication*). The fourth hypothesis does not face the question of why males should remain in close proximity, or even aggregate, such that special behavior to reduce acoustical interference should evolve.

In other words, these hypotheses were largely developed without the modern view that selection is principally effective at or below the individual level and that explanations invoking physiological limitations are open to question unless selective conflicts which force compromises are identified. The shortcomings of these hypotheses illustrate the magnitude of the revolution in evolutionary thinking that has occurred within the past decade and the value of this revolution in enabling us to develop appropriate models of the evolutionary background of phenomena like specialized chorusing in acoustical insects.

Walker (1969), Otte (1972, in press), Lloyd (1971, 1973*a*, *b*, in press), Otte and Loftus-Hills (unpublished), and Otte and Smiley (unpublished) have written more recently on aspects of chorusing behavior (or synchronous flashing), and all have endorsed versions of the fourth of the above hypotheses. The problem of why males aggregate or remain in close proximity is still unsolved, with only Otte (in press) considering it in some detail. Walker, discussing the snowy tree cricket, suggested that males in groups may be less vulnerable to acoustically orienting predators. Otte (1972) and Lloyd (1973*a*) were the first to emphasize in print that male-male phonoresponses are likely forms of competition for females, and Lloyd (1973*a*) and Otte and Smiley (unpublished) specifically suggested that mass synchronies are most likely incidental effects of such competition.

THE NATURE OF CHORUSES

Choruses of singing insects and flashing synchronies of fireflies sometimes reach astonishing proportions, impossible to describe adequately to someone who has not observed them directly. They are easily the most awesome performances in animal behavior that I have witnessed in the field. I will discuss a few examples briefly before trying to dissect or model their functions and reconstruct their evolutionary histories.

In early June 1957 after several consecutive days of unsynchronized chorusing during rainy weather, the males of one of the two seventeen-year cicada species I was studying (Alexander and Moore 1958) suddenly (within a few seconds) entered into a massive synchrony extending several hundred meters, from one end of the forest to the other. The only record that could be made of this observation was a tape recording of the sound (Alexander 1960) and a line graph showing the regularity of highs and lows in the intensity of sound as recorded by a sound-level meter (Alexander and Moore 1958, fig. 8).

In subsequent years I made two similar observations on other signaling insects, one of synchrony in a 400-meter-long population of fireflies (probably *Photinus pyralis*) in a river-bottom pasture in Kentucky, and the other of synchronized alternation in dense populations of true katydids in the Appalachian Mountains (Alexander 1960).

For a long time biologists have claimed on occasion to have observed phenomena like these. But most field biologists fail to notice events of such magnitude, and the rare reports may often be attributed to exaggeration or overenthusiasm. Thus mass synchrony in *P. pyralis* has been reported only rarely (see Buck 1935), and it has never been witnessed by Dr. James E. Lloyd of the University of Florida, who has surely spent more time than any other person studying firefly behavior in the field in eastern North America. Recently Otte and Smiley (unpublished) have analyzed mass synchrony of a related species, *P. concisus*, in Texas.

Similarly, in 1960 (p. 78) I described the synchrony of a dense population of the meadow katydid, *Orchelimum vulgare*, in a bed of tiger lilies on the campus of the Ohio State University. This species is one of the most widespread, abundant, and noticeable roadside singing insects in North America, but my description of its chorusing synchrony was novel. That description was also one of the items doubted by the reviewers of the publication, who had worked extensively with the same species in the laboratory. That particular population of long-winged immigrant males synchronized every day almost within hearing of my laboratory, and I have repeated the observation in the same location on two different years since 1960; but I have rarely observed synchronized choruses of this species elsewhere.

Even if such difficult-to-corroborate observations are

accepted, these extremes of chorusing behavior are achieved but rarely. As a result, we can legitimately doubt that truly *massive* synchronies, in these cases, have an evolved function. Instead, we are likely to speculate that they are incidental effects of a function that yields a somewhat less spectacular result and is realized more consistently.

On the other hand, Melanesian fireflies in the genus *Pteroptyx* regularly achieve massive synchrony (Lloyd 1973*a*, *b*), leaving us with the question whether or not they may be more than incidentally different from those which achieve synchrony only infrequently. Likewise, the synchronizing species of seventeen-year cicada (and its thirteen-year counterpart or cognate species)(Alexander and Moore 1962). regularly achieves and maintains synchrony within fairly large subpopulations even if it only rarely does so throughout an entire forest. Neither of the other two seventeen-year (nor the other two thirteen-year) cicadas seems to synchronize at all. Again, we must ask whether the frequently achieved smaller-scale synchrony of such species has an evolved function, or is an incidental effect of, say, a structural difference between the species' calls, or of a function not so dramatic in its results and shared with those species that never achieve synchrony.

Considering still smaller groups, we can note that, when two male insects that signal acoustically are placed near one another, they do not signal independently, each as if the other were not there. Instead, they interact, perhaps inevitably—see hypothesis (a) above (Alexander 1960)—and these interactions, termed *phonoresponding* (Busnel and Loher 1961), take certain forms that can be related to the chorusing behavior of larger groups. Thus some males synchronize calls with neighbors, some alternate calls, and some sing overlappingly or sequentially in ways that are not easily categorized as either synchrony or alternation (Alexander 1960, 1967; Dumortier 1963; Jones 1966; Shaw 1968; Otte, in press).

Even in interactions between two males, we can derive evidence that phonoresponses may occur as incidental effects. The coneheaded grasshopper, *Neoconocephalus ensiger*, is a common nocturnal singer along roadsides in northeastern North America. It produces a lispy pulse in series of indefinite length at rates varying from 4-5 per second at the lowest singing temperature of about 9°C to about 15 pulses per second at 30°C (Borror 1954). At high temperatures neighboring males do not synchronize; one can hear their songs go in and out of phase, even though the pulses come too rapidly to be counted.

This lack of synchrony is not surprising, since specialized chorusing behavior (synchrony and alternation) seems to be restricted to those species "in which the normal calling song contains a precise or highly uniform chirp or phrase rate within the range of two to five per second . . ." and to include all such species (Alexander 1960, p. 82). The pulses of *N. ensiger* are uniform in rate, but only at the lowest singing temperature are they delivered slowly enough to allow synchrony. At temperatures within a degree or two of the minimum at which this species will sing (approximately 8.6°C), I have on three occasions listened to males less than one meter apart synchronizing uninterruptedly for several minutes. It scarcely seems likely that these males have been selected to synchronize when the conditions under which they can do so effectively are rarely encountered. Rather, their songs appear to become synchronizable at very low temperatures as an incidental effect of their structure at more usual singing temperatures.

Additional support for synchrony or alternation as an incidental effect may be derived from its occurrence between individuals of different species (Alexander 1960; Littlejohn and Martin 1969). Alternation between sympatric species that sing together may be adaptive, but one can also obtain alternation regularly in the laboratory between species that are unlikely to interact in the field. In my laboratory a male of the katydid *N. exiliscanorus* (120 phrases/min) sang only between the phrases of a nearby male of *N. nebrascensis* (30 phrases/min), the two males thus effectively alternating with one another (see Alexander 1956 for song descriptions and audiospectrographs). Yet the males of these broadly sympatric species synchronize with conspecific males in the field.

Males may be selected to alternate with any sounds with frequencies near enough to their own to interfere significantly, or this particular case might have been an incidental result of the song difference between the species. If phonoresponses between near neighbors should prove inevitable, even if sometimes maladaptive, the conflict in selection that is responsible could lie in the value of auditory feedback to the maintenance of steady, rhythmic singing by lone individuals.

At the outset, then, it would appear that we can draw five tentative conclusions about the particular form of chorusing behavior called synchrony:

1. Synchrony is not always an evolved phenomenon, evidently appearing sometimes as an incidental effect of

close proximity of singing males having certain kinds of calls.

2. Synchrony occurs so regularly in some species, between neighboring males or throughout populations, as to suggest that its likelihood and scale have been enhanced, directly or indirectly, by natural selection.
3. The likelihood of synchrony is increased by close proximity; hence its expression throughout a population depends upon relatively high densities or tendencies to aggregate.
4. In at least some species (e.g., *Magicicada cassini, M. tredecassini, Orchelimum vulgare*) males are attracted into close proximity by the same signals that they synchronize once they have aggregated.
5. A noticeable effect of synchrony, to the human observer, is an emphasizing of the synchronized pattern in the song, which would otherwise be obscured increasingly as more males called while near one another.

Since males evidently must be close to one another to synchronize or alternate effectively, chorusing behavior raises two basic questions: (1) what advantages to individuals derive from synchrony or alternation? and (2) what advantages derive from closeness or aggregation of competing males?

For any case of specialized chorusing that is not simply an accident of close proximity, there are three possibilities.

1. Dense populations or aggregations have recurred so consistently and for such a long time that males have evolved to compete with their neighbors through phonoresponses that lead incidentally to chorusing (Lloyd 1973*a*; Otte and Smiley, unpublished).
2. Aggregating has been favored, with an adaptive significance to individuals that is enhanced by chorusing.
3. Chorusing has been favored by selection, leading to tendencies in males to enhance chorusing by aggregating.

These questions are part of the larger problem in biology of discovering how sexual selection and mate competition have led to various kinds of male signaling and breeding aggregations. What have been the roles of differential parental investment, environmental patchiness, and predation in determining the forms of these activities?

HOW CAN MALES COMPETE BY PHONORESPONDING?

Consider the dilemma of a male cricket or katydid calling near another male. He can respond to his competition by leaving the vicinity to try his luck elsewhere, but this may entail moving to a location that is more exposed to predators, farther from food, or farther from the females when they are newly sexually responsive. Moreover, he may have to travel a great distance to escape similar competition from other males.

He can attack the other male and try to drive him from the area, but this may be an expensive act involving time, energy, and the risk of damage in the fight or exposure to predators. It also requires temporary cessation of signaling, and it may be unsuccessful. There may be more than one such competitor, and displaced males may be quickly replaced, necessitating repetition of the whole expensive act, with every outcome doubtful. Evidently, there are limits to the benefits realized from aggression toward nearby competitors, or retreat from them, which restrict the circumstances in which these responses are expressed.

A third strategy, which may be employed whenever the expense, risk, or uncertainty of outcome of retreat or attack is too great, is to remain in the same location and outsignal the competitor. Among species with long-range signals, some expression of this strategy is likely to be universal, for males will rarely be able to signal in complete isolation.

To outsignal a nearby competitor, a male may, first, call more loudly or more frequently than his neighbor. We may assume that the resulting selective "races" would lead to the loud, persistent calling of modern crickets and katydids, and also to tendencies by males to direct or concentrate the broadcasting of their signals in the most favorable directions, to whatever extent and in whatever fashions those might be determinable (see Lloyd 1973*b* for evidence of directional signaling in fireflies).

Additionally, a male should gain from maximizing the fit between the pattern of his song and that favored by the females of his species. This he can do by (1) producing the appropriate pattern and (2) minimizing its loss of distinctiveness or its obscurement as a result of the singing of nearby competitors.

Suppose that the song of the species is a long trill composed of very rapidly delivered pulses, and the pulse rate (not the length or spacing of some group of pulses) is the chief pattern element of significance to the females (e.g., various *Oecanthus* species: Walker 1957). Aside from singing whenever his neighbors sing, a male in such a species has three possible strategies of phonoresponding: start first, stop last, or continue singing during periods when neighboring males are forced to stop. This behavior would lead to mutual stimulation into song by neighboring males, and to bouts of singing by groups of males; and it would lead to the evolution of long-continued songs that overlap one another. Such interactions in fact characterize species with songs of this type (Alexander 1960, p. 77; Otte 1972).

Now suppose instead that the song consists of series of short, multipulse chirps, with the pulse rate, pattern, or number within the chirp a key signal to the female (e.g., B_1 pattern: Alexander 1962). Zaretsky (1972) demonstrates the significance of pulse pattern in a song of this general type in the cricket *Scapsipedus marginatus*. To keep his own song maximally effective, a male in such a species should place his chirps between those of nearby competitors so that the critical within-chirp elements are in the least danger of being blurred. This tendency would lead to alternation of chirps, or a minimizing of signal overlap. Such behavior, again, is characteristic of species with songs known, or likely, to fall into this category—that is, songs with widely spaced chirps and slow or intermediate pulse rates (Alexander 1960, p. 79). It also occurs in hylid frogs with long, single-pulse, frequency-slurred chirps (Rosen and Lemon 1974).

Two kinds of alternation may be distinguished: (1) that in which the chirps or phrases of two males tend to be equidistant in series (e.g., the true katydid, *Pterophylla camellifolia*: Alexander 1960; Shaw 1968) and (2) that in which one male's chirps or phrases follow those of the other by a closer interval (e.g., *Hyla versicolor*: Rosen and Lemon 1974). Only in the former case, evidently rare, does mass alternation occur, giving the effect of a synchrony, in which male A alternates with male B, male B alternates with male C, and male C is thereby synchronized with A.

Finally, consider species in which the communicative unit or morpheme (at least at long range) is a pattern element repeated slowly, such as the entire chirp and its interval, irrespective of the fact that each chirp may be made up of several rapidly delivered pulses or wingstrokes (e.g., the snowy tree cricket: Walker 1957). In this case, neighboring males singing in alternation would deliver an abnormally fast song to a female far enough away to hear both of them. Only if two such males synchronized their chirps would the female be attracted to them, and then she might be attracted to them rather than to an equally distant single male because of the greater intensity of their combined songs. Again, males in species with such songs seem inevitably to synchronize with their neighbors (Alexander 1960, p. 77-78; Walker 1969). Otte (in press) reports an observation of a South American treetop species with a snowy-tree-cricket type of song alternating, but he could not determine the pulse rate,[1] and it seems likely that this was instead one of the chirping eneopterines with the B_1 pattern in which chirps are alternated.

Otte and Loftus-Hills (unpublished) note that an additional strategy may be to interfere with the song pattern or rhythm of a neighbor to prevent females from being attracted to it. Such "spiteful" behavior (Hamilton 1970) would obfuscate the rhythms of both singers, and would have to be more beneficial to the individual practicing it than alternatives yielding more direct benefits, such as keeping his own rhythm separate and clear. Hence it seems relatively unlikely to evolve.

The above analysis includes the three major classes of phonoresponses (Alexander 1960, 1967) and the major categories of simple song patterns (Alexander 1962), and it seems to predict accurately the phonoresponses of acoustical insects on the basis of the rates and patterns of delivery of the morphemes, or communicative units, in their songs. It also leads to the tentative conclusion that the phonoresponses that form the bases of chorusing behavior are in fact means of signaling competition between neighboring males. The question that remains, and which we will defer temporarily, is: Do males ever gain by aggregating because group synchrony is thereby promoted, or do they aggregate for other reasons, with the group synchrony either an incidental effect or an enhancer of the function of aggregation?

MALE COMPETITION AND SIGNALING/SEARCHING TIME

An additional competitive strategy is possible in aggregations of singing males. At low or moderate densities, males unable to secure suitable (e.g., predator safe) signaling locations may obtain matings by lurking in the vicinity of calling males and intercepting responsive females on their way to the callers.

---

[1]Personal communication.

This strategy should be least profitable (1) when callers are so far apart that only one male can be cuckolded at a time and (2) when there are no decidedly superior males or singing locations. It should be profitable (1) at moderate densities when a nonsinger can effectively patrol the peripheries of the territories of several signalers, (2) when calling males are very unequal in signaling prowess, and (3) when resources are clumped and one or a few males are able to retain decidedly superior signaling sites. The last two conditions are likely to be maximized for some anurans in which adult males continue to grow and develop better voices and fighting ability for several years, and in which breeding sites such as ponds or streams frequently contain but one or a few superior singing sites. Restricted directions of approach by females will further increase the likelihood of cuckoldry by silent males searching near a signaling male.

It seems apparent that as densities increase males will profit from increasing the proportion of their time spent (a) patrolling for interloping males or (b) searching for approaching females to reduce the likelihood of cuckoldry (fig. 1). At high densities males may sometimes profit from searching without signaling, even if males and signaling sites are not decidedly uneven in value. Only one such case has been suggested among vegetation-inhabiting acoustical insects (Otte and Joern, in press)—the grasshopper, *Goniatrum planum*—and it is apparently associated with the presence of relatively few singing sites (bushes). Lloyd (1973*b*) also notes that many nonsignaling individuals occur in swarm trees of *Pteroptyx*.

On the other hand, parasitism of callers by small, silent (usually younger) males may be common among anurans (Axtell 1958; Emlen 1968)[1] and probably occurs among surface-dwelling crickets[2] and grasshoppers.[3] Surreptitious matings on the periphery of harems, and other alternative routes to mating success by males, are described by Le Boeuf (1974) in the highly polygynous elephant seal. Probably, the peculiar satellite males of the ruff (Hogan-Warburg 1966) can be interpreted as an extreme case in which the evolution of dimorphism in male behavior has led to a divergence in plumage as well. Gadgil (1972) develops a theoretical argument for this situation.

---

[1]Richard Howard: personal communication.

[2]William Cade: personal communication.

[3]Daniel Otte: personal communication.

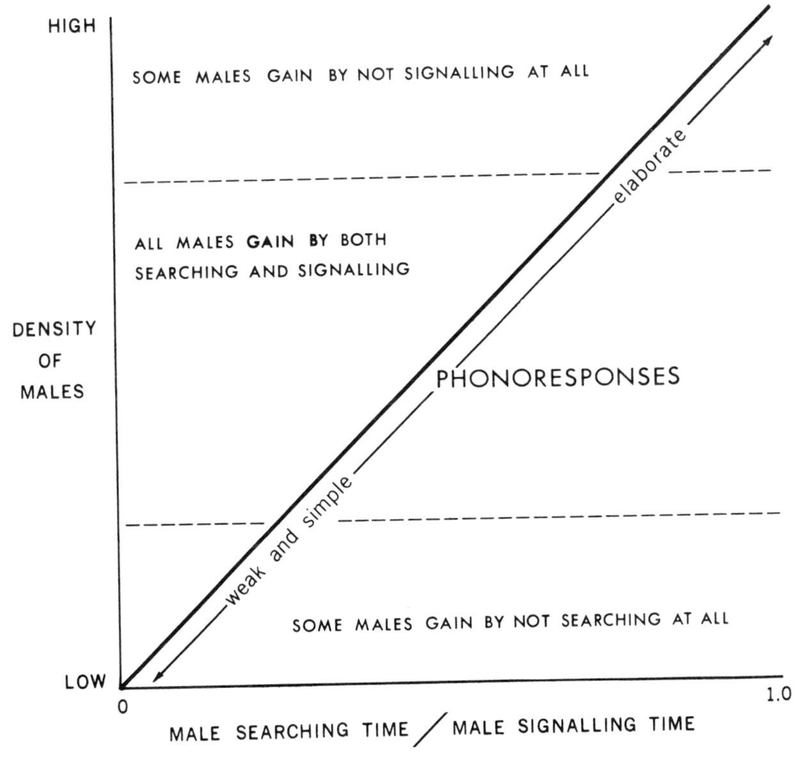

Fig. 1. Postulated relationships between density of males of acoustical insects and proportions of time spent signaling and searching. Densities may be increased by active aggregation. Signaling/searching time will also be affected by signaling ability of males, quality of signaling sites, and distribution of females.

Among male-dimorphic insect species, the possibility must also be considered that the two kinds of mating situations resulting in two male morphs may not be coincident in time or space, and there may be no direct sexual competition between the two kinds of males. Thus, in species living in temporary habitats, some males may gain by being relatively aggressive and territorial because they mature in established but still desirable habitat, while others gain from being better fliers and searchers because they mature in deteriorating habitat.

Failure to find evidence of nonsinging males searching through cicada choruses implies that the male's song remains an essential part of his ability to acquire females. This hypothesis is strengthened by the similarity of the initial courtship song to the calling song (Alexander 1968) and by observations (Alexander and Howard, unpublished data) that the time from landing by a chorusing male to the start of his next song is shorter and more uniform in length than the time from termination of a song until the subsequent flight. These data indicate that incoming females at some point orient on the songs of individual males.

The periodical cicadas may provide a test of the effects of varying densities on proportions of time spent by males in searching and signaling, if the flights of male cicadas between songs can accurately be interpreted as reflecting time spent searching for females (fig. 2). The three *Magicicada* species with each of the two life-cycle lengths (Alexander and Moore 1962) differ rather consistently in population density, and their calls and behavior are somewhat different. The two-*decula* cognate species are usually sparse and have an obligatorily long call, which may be interpreted as a relatively low searching-to-signaling ratio if flights between calls do not differ greatly in length from those of other species. The other two pairs of species are commonly abundant and have short calls which they produce in series of two to three between flights. However, when most dense, males in both of these pairs of species appear to reduce the number of calls between flights, implying an increase in the searching-to-signaling ratio. When densities are very high and weather conditions are optimal, the -*cassini* cognates reduce calls between flights consistently to one and synchronize their calls. As might be predicted from these species differences, males of the -*cassini* cognates are sexually more aggressive and more successful at interspecific matings (Alexander and Moore 1962). The observations leading to these correlations (Alexander and Moore 1958, 1962; Alexander 1969) were carried out in the absence of the hypothesis they seem here to support.

OTHER COMPETITIVE STRATEGIES

Competitive strategies of males signaling in groups, in addition to the general categories discussed above, include the following:

1. Seek out other males that appear to be courting or to have located a female (perhaps even neighboring males that have simply stopped singing) and attack the male

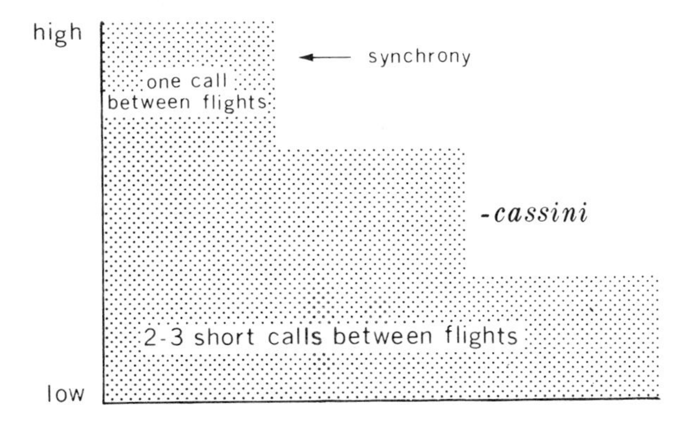

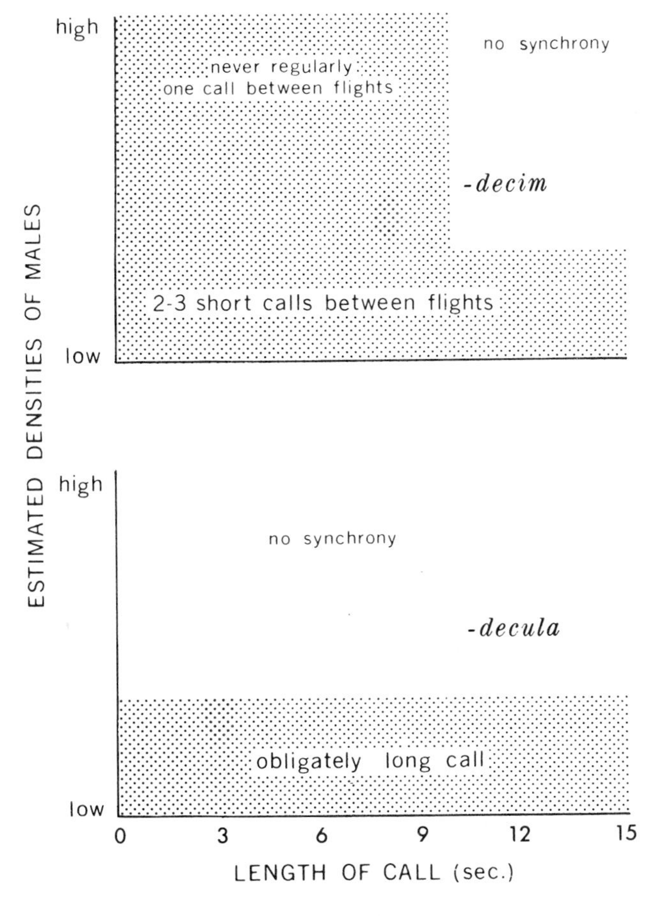

Fig. 2. See legend on next page.

or try to break up the interaction or steal the female. Alexander (1961) reported that a male of a *Gryllus* species, for example, repeatedly searched out and attacked another male out of sight in the same container whenever the latter began to court. Otte (in press) describes similar behavior in a grasshopper. Grove (1959), Spooner (1968), and Lloyd (1973*b*) describe behavior with the apparent function of usurping the place of another male exchanging signals with a responsive female.

2. Reduce signal intensity when females are near and the likelihood of cuckoldry is high, or change to a less risky channel (e.g., from sound to vision or substrate vibration) (see below and Lloyd 1973*b*).
3. Rather than ceasing song when it is particularly dangerous (e.g., daytime) or calorically expensive, reduce the intensity of the song or the rate of calling (see below).
4. Change to a pattern, rate, or channel with more directional information or more precision when females are nearby or likely to be nearby (see below).

The last of these strategies should be followed when females are consistently being attracted at such long range that they are using some simpler unit to determine that the male is a proper one to approach (see also Lloyd 1973*b*). At long range, for example, females of the snowy tree cricket are unlikely to detect the pulse rate or pattern within chirps if several males are synchronized, since synchrony is not exact. Females at long range probably respond chiefly to the chirp pattern and the frequency of the sound. Snowy tree crickets deliver the pulses within their chirps in two or three groups, raising the question whether the pulse pattern within chirps represents a hierarchy of patterns maximally effective at different distances in a synchronized chorus. Whether or not this interpretation is correct, I suggest that shifts in the critical signal parameters with distance from the signaler will prove to be common among acoustical insects.

---

*Fig. 2. Evident relationships between density of males and alternation of calls and short flights in males of the three pairs of cognate species of periodical cicadas (*Magicicada *spp.). Densities of males in choruses are enhanced by active aggregation in the* -cassini *and* -decim *cognates (Alexander and Moore 1958), and probably also in the* -decula *cognates.*

INTRASPECIFIC COMPETITION AND COMPLEX SIGNALS

Spooner (1968), Lloyd (1973*b*), and Otte (1972, in press) describe the functions of complex repertoires in katydids, fireflies, and grasshoppers, respectively, in signal systems involving alternating responses between males and females during rapprochement. Evidently, all four of the strategies described above are involved in these interactions, with males changing the intensity, pattern, or communicative channels of their signals as females approach, and with other males carrying out competitive or usurping behavior. In these cases, males can respond directly to approaching females because the females signal, and other males can interlope for the same reason. These cases may provide clues toward the difficult problem of explaining complex signals in species in which the females do not signal back to calling males.

Pace (1974) noted that the song of the male northern leopard frog, *Rana pipiens*, to which females do not respond acoustically, is composed of three main kinds of phrases most often produced in a particular sequence, A-B-C. If A sounds are played at a group of silent males, they respond with increased proportions of B and C sounds. A similar observation has been made on the cicada *M. cassini* (Alexander and Moore 1958) and on certain meadow grasshoppers with two-part songs (Alexander 1960).

Pace and I noted in the field that from a great distance humans, at least, are likely to hear only the A sound of *R. pipiens*, which is a long gutteral trill (Pace 1974). The other sounds are easily heard only at fairly short range, and they are sometimes missing from the repertoire of males of the related species, *R. palustris*.[1] Accordingly, Pace speculated that the B and C sounds of *R. pipiens* are, respectively, close-range, direction-giving and male-male, aggressive signals. Whether or not this is precisely correct, the important point is that such changes in intensity or pattern may represent specializations for long-range attraction and short-range direction giving, or alterations, such as in intensity, that specifically reduce the likelihood of cuckoldry.

In frogs, approaching females may be detected by swimming sounds; splashing causes approach and clasping in some species (Eibl-Eibesfeldt 1954) and production of the "mating

---

[1]A.E. Pace: personal communication

call" in *R. utricularia*.[1] Some katydid males respond to agitation of the plant on which they are calling (as by an approaching female) by vigorously shaking or vibrating the plant (Busnel, Pasquinelly, and Dumortier 1955), which may be an aggressive signal commonly directed against other males, but is also a signal potentially transmissible to the female but not to nearby males on other plants. Such a signal would result in less cuckoldry than an acoustical signal that other males could use to discern that a female was approaching the signaling male. It seems a logical extension that call changes may evolve not only to occur when acoustical responses or other evidences of approaching females are detected by calling males, but also in some cases to be regularly interspersed or alternated with long-range calls, even when the male has received no direct evidence of approaching females.

The kinds of selection discussed above could have several different effects on a male's singing; these effects may relate to the variety of complex calling known in acoustical insects in six ways.

First, long trills may change progressively, as in *Amblycorypha rotundifolia* (Alexander 1960). In this species trills tend to increase in length during a singing bout, and each trill series ends with one very long trill followed by a few short trills.

Second, long trills may alternate between two speeds or intensities, as in *Conocephalus strictus*, *Tibicen* species (Alexander 1956), and *Metrioptera sphagnorum* (Morris 1970; Morris, Kerr, and Gwynne, in press). Alternation of slow and fast pulse rates in the trill of *C. strictus* suggests that at long distances females may respond chiefly to the frequency spectrum of the sound, while at short range a distinctive pulse rate becomes more important, favoring a slower pulse rate with each pulse made distinct. In several cricket species, such as *Allonemobius allardi*, *A. tinnulus* (Alexander and Thomas 1959), and *A. griseus* (Alexander, unpublished data) the pulse rate in the courtship sound is slower than that of the calling song.

In *Tibicen* species loud and soft buzzes alternate, with the loud portions in many species containing a slower vibrato superimposed on the basic trill (Alexander 1956). The presence of the louder, slower vibrato implies that at longer ranges the rapid pulse rate in the buzz is not easily discerned. The soft buzzes may represent an advantage over

[1] A.E. Pace: personal communication.

silence during the same periods because they attract nearby females. Similarly, the so-called daytime songs of *Scudderia* species (Spooner 1968) could be low-effort signals which profit a male only if females are nearby, but are also less likely to attract birds than normal nocturnal calling.

Morris, Kerr, and Gwynne (in press) have shown that the two parts of the song of *Metrioptera sphagnorum*, alternated twice each second, and termed by them the *ultrasonic* and *audio* modes, are both functional in pair formation, but "the ultrasonic plays the primary role in pair formation. . . . A courtship song, apparently derived from the audio mode, is given by the male after pair formation is achieved." This finding strongly supports the hypothesis presented here that long- and short-range signals may be alternated regularly in the singing of a male not yet in contact with a female.

Third, some species seem to have two distinct calling songs, as do *Microcentrum rhombifolium* (Alexander 1960) and a Fijian species of Nemobiinae (Alexander and Otte, unpublished). In some *Scudderia* species females respond acoustically to one of the male's calls, causing the male to change to another calling sound (Spooner 1968). Males of *M. rhombifolium* alternate two calls (lisps and ticks) without acoustical responses from females. Results obtained by Grove (1959) and Spooner (1968) indicate that females move toward low-intensity male lisps but not toward high-intensity lisps. Females tick in reply to male ticks, and males then go to the female. Evidently, the lisp (a very rapid series of toothstrikes) is a long-range signal, and the tick series (approximately the same series of toothstrikes produced very slowly) is a short-range signal used by the male to locate the female.

As with *Scudderia* species, a *M. rhombifolium* male apparently mimics the female's response to his ticks, which probably confuses potential cuckolders (Spooner 1968; Grove 1959; Alexander 1960, recording); Grove (1959) has observed otherwise silent interloping males producing this sound while approaching a phonoresponding pair.

The Fijian nemobiine may be similar. Males have been seen to change from one call to the other without touching other individuals. Females, however, are not known to call. The two calls of this species differ most dramatically in wingstroke rate; both are composed of short, multipulse chirps.

Fourth, each phrase or chirp may change progressively, as in *Orchelimum volantum* (Alexander 1960) and *Magicicada* species (Alexander 1968). In these species the pulse rate slows at the end of the phrase, and the frequency of the sound drops. In all *Magicicada* species the initial courtship sounds consist of phrases resembling the calling phrases. In the *-cassini* and particularly the *-decim* cognate species pairs, the initial courtship sounds are shorter than calling phrases, and separated by shorter intervals, so that in effect the terminal part of the phrase is repeated rapidly. This part of the phrase may thus be principally a short-range signal with a function near that of the courtship signals. In at least one cricket, *Allonemobius fasciatus* (Alexander and Thomas 1959), a progressive change at the end of the chirp, involving a slowing pulse rate and a lowering frequency, appears only in the courtship song.

Fifth, phrases or chirps may have a distinct two-part pattern, as in most *Orchelimum* and *Conocephalus* species, *Magicicada -cassini* cognates and *Teleogryllus* species (Alexander 1956, 1962, 1968; Zaretsky 1972; Hill 1974). These songs can be compared with the two-part trills described above. Some *Teleogryllus* species are somewhat intermediate in that their phrases are repeated rapidly without noticeable intervals, so as to sound like continuous two-part trills. Males of this group tend to answer one part of their two-part phrase with the other part (Alexander 1960). The *M. -cassini* cognates seem to reverse the order of the two parts of their calling phrase in the initial stage of courtship, and to reduce the length of the silent intervals, as do the *M. -decim* cognates.

Sixth, there may be a complex multipart song, as in *Amblycorypha uhleri* and its (undescribed) siblings (Alexander 1960; Walker and Dew 1972; Walker, unpublished) and *Scudderia septentrionalis* (Alexander 1956). The song of *A. uhleri* is an astonishing output for a signaling insect. While *Amblycorypha* females do respond acoustically to males, lone males of *uhleri* nevertheless repeat a programlike sound lasting 45-60 seconds each time they sing, its chief variables being the number of repetitions of certain parts (Walker, unpublished).

In light of the above analysis, the parts of this complex song may be more understandable. In its early phase the song consists of a two-part trill, the first part longer and with a faster pulse rate. This is followed by ticks and a series of phrases, each phrase slowing in pulse rate and dropping in frequency and intensity, as in the calling phrases of *Magicicada* species and *Orchelimum volantum* and the courtship sounds of *Allonemobius fasciatus*. *Amblycorypha* females, like those of several Tettigoniidae (Spooner 1968), seem to fly while initially approaching males, and males move frequently between songs. Perhaps the males in some such species have evolved to deliver a series of sounds normally sufficing to attract step by step through much or all of her approach any responsive female near enough when the male begins song to reach him (or signal to him) before he flies to another singing location. Supporting this hypothesis are observations by Thomas J. Walker[1] that females are most likely to phonorespond acoustically to the second (slow) trill in the song, especially if the slow trill is preceded by the fast trill (evidently the long-range signal). The third and fourth parts of the song of *uhleri*, its ticks and gradually changing phrases, are produced during the male-female rapprochement (which is not yet completely understood), and variations in the numbers of their repetitions are chiefly responsible for variations in the length of a male's song.

Some recently published results with frogs are perhaps relevant to the above hypotheses and observations. Oldham (1974) found that two females of *Rana utricularia* (Oldham used the name *sphenocephala*, but see Pace 1974) approached only to within 60-65 cm of a recording of the "mating song." Perhaps, as suggested by Pace (1974) for *pipiens*, additional signals, which also characterize *utricularia* calls, are needed for closer approach. Littlejohn and Watson (1974), in a study of the responses of *Crinia* species to acoustical signals, suggested that "the introductory note in the mating call of *C. victoriana* . . . may be used for directionality . . . the more distinctive repeated notes [providing] . . . the temporal coding for specificity of response." The possibility that gross elements of rhythm, and perhaps frequency, may be more significant at longer ranges, with finer or at least different aspects of pattern more important at closer range, is a problem that must be taken into account in designing and assessing the results of experiments with phonotaxis. An additional problem in such experiments is that females of some species are probably sexually responsive only in the presence of several or even a large number of signaling males (see below).

Complex or multipart songs, then, may be expected to evolve (1) when males have reliable ways of discerning that responsive females are nearby (as when females signal acoustically or approach noisily) or are likely to be nearby (as in chorusing aggregations); (2) when likelihood of cuckoldry is high (as when the male-female rapprochement is mediated by alternating responses); and (3) when the signal structures optimal for long- and short-distance attraction of females are different. Interspecific interference leading to character displacement remains a possibility (Alexander 1969;

[1]Personal communication.

Walker 1974), as does the alternation of signals significant to the female and to the male (Alexander 1960; Spooner 1968; Morris 1970).

CHORUSING AGGREGATIONS AND COURTSHIP SIGNALS

Alexander (1962) noted that the long-range, pair-forming signals of crickets and their relatives must have evolved from short-range, simpler sounds, most likely functioning in a courtship context. This plausible argument nevertheless leaves several questions unanswered. First, what exactly is a "courtship context," and is it the same in all species? Second, why do so many crickets and grasshoppers have complex and distinctive acoustical courtship signals, while most katydids and cicadas lack them? Third, under what circumstances might courtship (short range) sounds evolve secondarily from calling (long range) songs (suggested for some nemobiine crickets by Alexander 1962)?

Alexander and Otte (1967*b*) discussed a cricket species, *Hapithus agitator*, in which the long-range calling sound has evidently been lost in part of the species' domain, leaving only a soft sound produced at short range and described as a courtship sound. We have since (Alexander and Otte, unpublished) located several species lacking calling songs which, like *H. agitator*, tend to live in sedentary colonies or clusters, often on certain plant species (unlike most crickets with long-range acoustical signals).

These observations imply that, when crickets begin to live in dense clusters, often associated with a narrowing range of host plants, they tend to lose their long-range signals. Presumably, this happens when searching without signaling becomes profitable for most males. This point may be reached when the species starts to live on one or a few patchily distributed plant species, when odors or sights rather than sounds become most effective at long range, or when the adult females become easier to locate by searching out oviposition, hibernating, or other sites.

Species that live on the ground communicate in a two-dimensional environment, tend to have more distinctive courtship and aggressive sounds, are more likely to be diurnal than those living in vegetation (except for species that fly readily, as with cicadas), and do not synchronize or alternate in large choruses. The prevalence of aggressive sounds can be explained partly as a correlate of burrow or territory ownership; and the diurnal behavior, as a correlate of greater protection from predators, including burrows and retreats beneath surface litter, leaves, stones, logs, and so on.

The prevalence of courtship sounds is more difficult to explain. As contrasted with equally complex courtship involving chemical, visual, or tactual signals, acoustical courtship may also correlate with presence of retreats from predators, both because a courting surface-dweller usually has an escape location and because the importance of the retreat may make him more sedentary at closer range. (Such retreats may also be resources important to the females.) Males on vegetation are not so bound to special locations, hence perhaps can afford to pursue females during courtship.

Among species carrying out sexual behavior on vegetation, elaborate acoustical courtship signals have been known only in certain cicadas, namely the seventeen-year and thirteen-year species, in which the males compete in huge, dense choruses (Alexander 1968); their apparent absence in other cicadas, the above arguments on the evolution of complex signals, and the resemblance of the initial courtship sounds to calling all suggest that they have evolved secondarily.

The evolutionary order of appearance and loss of different kinds of signals provided by these observations and speculations is: (1) short-range signals (probably courtship) (numerous beetles and other insects may still be in this evolutionary stage: Alexander 1967), (2) long-range signals (pair forming) (most modern acoustical Orthoptera are in this stage), (3) loss of courtship in many species (e.g., most cicadas), (4) aggregation of signalers in a few species (for reasons not yet clear), (5) secondary evolution of courtship signals in some cases, (6) loss of pair-forming signals in a few of the aggregated species, and (7) loss of all acoustical signals in still fewer of the aggregated species.

These suggestions lead to the further implication that the long-range signals of insects did not arise as systems whereby groups of males could improve their ability to attract females from outside the group. Instead, they imply that the long-range signals of acoustical insects, today produced by both widely spaced and aggregated males, evolved in nonaggregating species, and that the leklike behavior of chorusing males in some species today is a secondary phenomenon. This hypothesis is supported by the rarity of acoustical species that aggregate clearly to sing.

It is worth noting here that mass chorusing is known only among vegetation-inhabiting species, and grasshoppers that crepitate in flight, although surface-dwelling species exhibit phonoresponses (Alexander 1960, 1967). This still unexplained correlation may relate partly to the importance of densities and distribution of signalers in chorusing behavior. A chorus on vegetation is three-dimensional, while a group on the surface is two-dimensional, with peripheral members close to few other individuals. An approaching female or male on vegetation or in flight can receive simultaneously the signals of numerous males in a three-dimensional (or two dimensional) chorus, while a walking individual approaching a signaling group on the surface will be much more limited in the number of signalers it can hear at once.

WHY DO SINGING MALES AGGREGATE?

When Fisher (1958) analyzed sex-ratio selection as an aspect of parental investment, he pointed up the necessity of weighing in the obligate cost of every act or trait, in caloric expenditure as well as risk of death or disablement, in calculating its net benefit (see also Williams 1966*b*; Trivers 1972, 1974). For two reasons, this view of the selection process has an immense value for the study of social behavior. First, it focuses attention on the altruism of acts that raise the fitness of others more than that of the actor, and on the difficulty of the concept of acts neutral to reproduction. Second, it reminds us that group living or aggregation entails certain universal expenses to the individual which must in every group-living species be overcompensated by advantages, none of which is universal (Alexander 1974).

Two disadvantages of grouping seem universal: increased competition for resources and increased likelihood of disease and parasite transmission. Perhaps no others are universal, although some may be frequent. For example, groups sometimes draw the attention of predators.

Most of the benefits assumed in the past for group living are wrongly based upon group effects detrimental to individuals. Elsewhere (Alexander 1974), I have argued that there may be only three categories of benefits from forming groups: (1) predator protection, either because of group defense or because the opportunity is afforded to place some other individual's body nearer the predator; (2) nutritional gains when utilizing food, such as large game, difficult to capture individually or clumped food difficult to locate; and (3) use of clumped resources (see fig. 3). Students have noted that two cases of grouping are difficult to fit into these categories: (a) communal winter clusters of flying squirrels, which may chiefly gain from reducing temperature loss (Muul 1968); and (b) the V-formation of migrating waterfowl, which may gain from combining their information about the migratory route (Hamilton 1968). Biologists studying mating aggregations, and social biologists in general, have not considered the severity of the problem of identifying the advantages of grouping behavior to individuals.

This brings us to the second major question relevant to this essay. Males of acoustical insects intensified their competition for females as their signals evolved to be effective at increasing distances. We can easily understand that there might be distances beyond which it was unprofitable for males to disperse to avoid singing near competitors. But how have males come to aggregate during signaling, as they do in some species of katydids, cicadas, and fireflies? Such aggregation, I believe, necessarily increases the intensity of competition for females once they have been attracted to the aggregation.

KINDS OF MATING ASSEMBLIES

There are two major classes of mating assemblies. First are the clusters of males competing to mate with females where the females are already aggregated, or are likely to aggregate: emergence sites, oviposition sites, feeding sites, hibernacula, and so on. Excellent examples are (1) the scatophagous flies (*Scatophaga* spp.), studied extensively by Parker (1970) in England, whose males compete violently for females at the dungpats where oviposition occurs; (2) the brown paper wasp (*Polistes fuscatus*), which mates on the nest or near hibernacula (West 1969); and (3) cicada-killer wasps (*Sphecius speciosa*), whose males emerge first and set up territories where emergence holes are most dense (Lin 1963). These are not necessarily signaling groups of males, nor are they easily perceived as direct precursors of signaling male aggregations (the situation most difficult to explain), since they seem more closely linked to female signaling. They are clearly examples of males competing for already aggregated females or controlling areas containing resources valuable to the females.

The second kind of mating assembly, which remains puzzling, is that (such as the periodical cicadas and Melanesian fireflies) in which the signaling males definitely

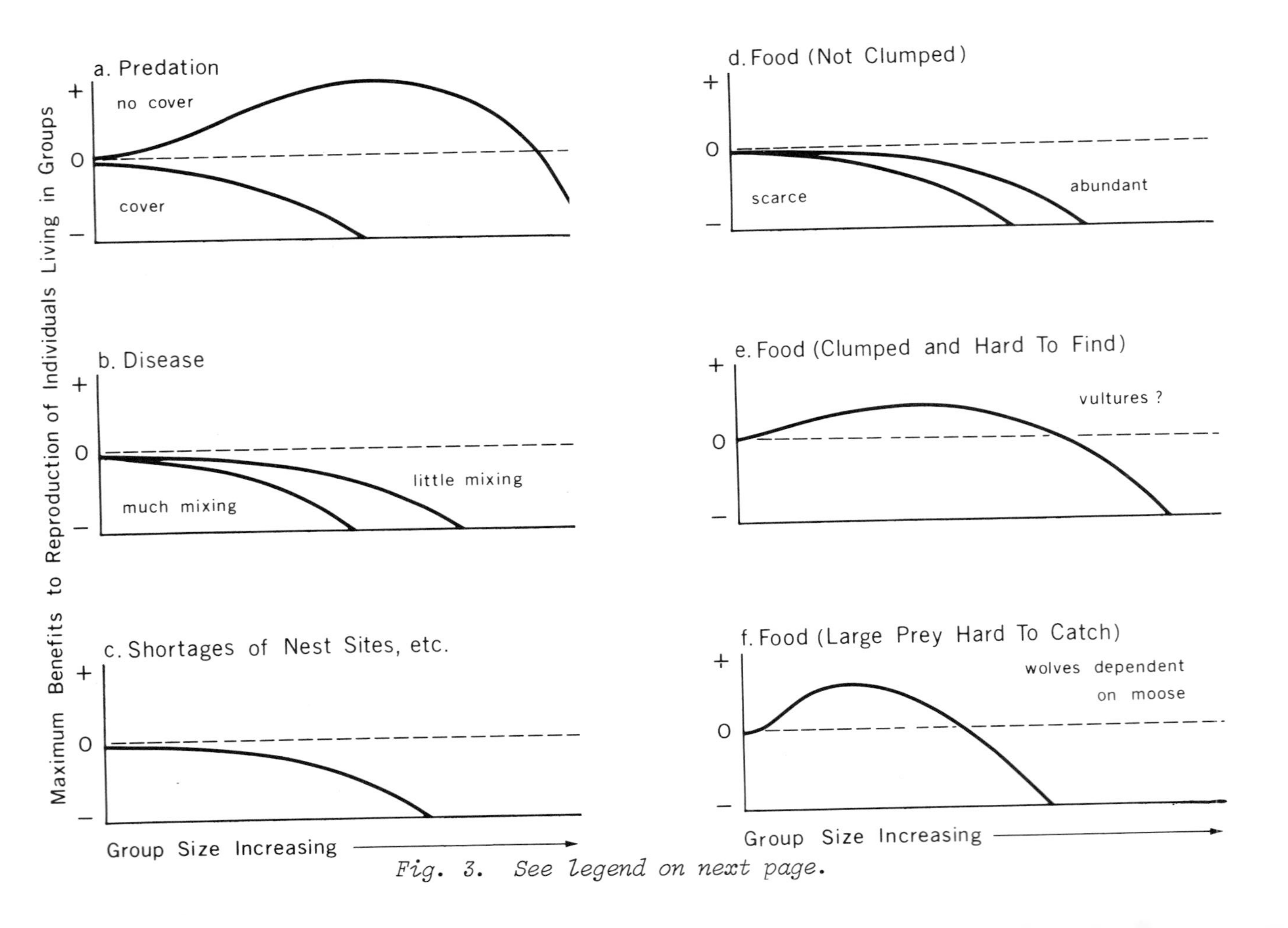

Fig. 3. *See legend on next page.*

aggregate (whether or not they synchronize or alternate on chorus), but not necessarily where the largest number of females is expected to occur, and which appears to attract large numbers of females from outside the group. In some cases, such as the snowy tree cricket in which males regularly synchronize or alternate in chorus, it is not clear whether or not males actually aggregate or whether or not they regularly attract females from outside the chorus.

What are the males in these cases doing? In some species males may in fact be aggregating where the females are likely to be, but that is not obvious to us. For example, Spooner (1968) found that males of the katydid, *Inscudderia strigata*, aggregate on their food plant, *Hypericum strigata*. Marianne N. Feaver[1] has evidence that males of the synchronizing katydid, *Orchelimum vulgare*, aggregate and sing in the vicinity of clumps of those plant species in the stems of which the females are most likely to oviposit. Morris (1972) has shown that males of *O. vulgare* are attracted to the songs of other males. The last two observations together suggest that both females and males of *O. vulgare* probably use male songs (as well as other cues) to locate appropriate breeding sites. Even if the females are attracted principally by the songs of the males, rather than by the presence of the appropriate plants, the situation is easy to understand.

The mating assemblies of periodical cicadas and Melanesian fireflies, as well as those of some dipterans and other insects (Downes 1958, 1969), take on the appearance of "lek," or "arena," behavior in various vertebrates. Lek-breeding vertebrates include species such as the Uganda kob (Leuthold 1966; Buechner and Roth 1974), the sage grouse (Patterson 1952; Wiley 1973), the European ruff (Hogan-Warburg 1966), and many others in which the males form groups and signal visually, acoustically, or both ways in special arenas, or courts, which are apart from the nesting or feeding grounds and do not seem to represent locales where the females are expected to aggregate for reasons other than mating itself (Lack 1968). In other words, the females come to these groups of males as they do to the choruses of the males in at

[1]Personal communication.

Fig. 3. Benefits to the reproduction of individuals from living in groups. Each curve is intended to approximate the maximum benefits likely from each context of group living, the actual benefits in any case representing a summation of the effects of predation, disease, and shortages of nest sites, food, and other resources. See also Alexander (1974).

least periodical cicadas.

Explaining insect choruses and explaining lek-breeding behavior seem to be part of the same general problem. Much has been written about vertebrate leks, but there seems to be no general theory of their origins. Nevertheless, valuable comparisons can be made between the reflections of vertebrate biologists and those of entomologists on this subject, and I shall use them both in this discussion.

THEORIES OF LEK-BREEDING BEHAVIOR

Lack (1968) stated that "communal leks" have evolved nine times in birds. In all cases, the males signal elaborately, most with exaggerated postures, some with bright plumage, and some principally with calls. Bird leks occur at traditional places used generation after generation by long-lived birds that presumably remember them. The only known case of similar behavior in insects is the recurrence of swarms of honeybee drones year after year in the same locations (Zmarlicki and Morse 1963; Strang 1970). This case is a puzzle, for no drones survive winter, and the queens that come to the swarms are also young virgins on their first extended flights. It is unlikely that chemical markers could be effective year after year. The most plausible explanation, based on data in the above publications, is that certain kinds of places, such as "clearings among trees," "slight summits" in open areas, or "areas marked by some form of vertical relief, whether it be forest, single trees, or buildings" (Strang 1970) are selected by both drones and queens, with drones and queens both also attracted by the presence of drones. But an explanation of the kinds of places selected for such aggregations does not explain the aggregations themselves.

Lack (1968, p. 155-56) supposes that arenas of bird leks are reused because they previously proved safe from predators. He notes that the behavior of the performing males of a lek appear "in all the species concerned" to be "adapted to reduce the risk of predation so far as possible." He even suggests that early detection of approaching predators, "may, indeed, be the main advantage of communal display," adding that "it perhaps helps the females to find males in breeding condition. . . ." He also suggests that "the final stage, from solitary to communal display, would follow if the males tend to display within earshot of each other, and if a clear grouping increases the chance, either of their detecting predators or of each male acquiring a mate."

Lack's predator explanation is important to us because it was taken up by Spieth (1974) to explain apparent lek behavior in Hawaiian *Drosophila*, and because Walker (1969) implied a similar explanation for the synchronized choruses of snowy tree crickets. If predation is invoked as the principal or primary cause of aggregations of males at mating time, then it will be difficult to suppose an historical relationship between lekking behavior and the widespread phenomenon of aggregations of males competing for females at emergence, oviposition, nesting, feeding, or overwintering sites. Yet it seems difficult to deny such a relationship.

Perhaps predation is invoked as an explanation for "true" leks because no other explanation has been apparent. But the presence of obvious adaptations to reduce predation may only be evidence that leks are particularly vulnerable to predation. One might as well argue that, because flocks of birds and herds of ungulates are almost always seen feeding, they must have originally formed their groups in order to feed more effectively. Unfortunately, this is the argument most prevalent in the literature, probably because feeding is obvious while predation in most cases is not (Alexander 1974).

Furthermore, for a long time it was not acknowledged that special advantages to individuals compensating the automatic disadvantages must always be present to account for grouping. So long as this requirement was waived, it was possible to interpret as a "feeding advantage" nothing more than a greater efficiency by the group in utilizing the food available in a particular area even if some individuals suffered; no one felt compelled to explain why these individuals didn't leave the group. In a similar fashion, Leuthold (1966) suggested that the leks of Uganda kob "offer certain advantages, such as providing a social organization and a spacing mechanism to the population and ensuring maximum efficiency of reproduction." Wynne-Edwards (1962) saw leks as "a method of controlling density of breeding birds." These explanations describe advantages to the population without explaining how the individuals of the population came to behave as they do (see also Lloyd's 1973*a* and *b* evaluations of similar arguments for firefly aggregations).

Brown (1964) used the sage grouse as a principal example in an effort to give a general explanation of leks; his explanation is similar to that advanced by Crook (1965). Brown stressed that for the sage grouse (from the work of Patterson 1952; see also Wiley 1973) food does not appear to be a critical factor in breeding behavior. There is no parental feeding of offspring or evidence of nutritional deficiencies in either adults or young during the breeding period, and nearly all of the mortality of the young seems to result from predation. Thus not only is the male evidently unimportant as a parent, his absence from the nesting area "is advantageous by decreasing conspicuousness of the family, and by reducing the potential prey population there (even if he were protectively colored)." Brown goes on to say:

> Thus freed from the responsibilities of protection and care of the nest and young the males have full freedom of competition for the fertilization of females. To this end have evolved the elaborate and conspicuous plumage and display in the males and the lek system of mate selection. Once evolved, the lek system tends to perpetuate itself through the demonstrated preferential success of the dominant males within the lek.

Brown's explanation does not face the question of how the individual males gain by initiating lek behavior. It does not explain, therefore, how tendencies to aggregate actually evolved. One wonders again why the males do not search out females individually rather than join a group of competitors who presumably will strive against one another to obtain whatever females may be attracted to the group.

Both Brown and Crook stress the value to the females of lekking birds of having the ruckus of intermale competition in the polygynous breeding system take place, with a minimum of risk, time, and effort, away from her nesting ground. There seem to be two ways by which this might have come about. Males may have originally held territories on the nesting grounds of the females. As long as the males were providing some benefits other than genes to the females, or at least were not affecting the females' breeding in a deleterious fashion by their presence and their competition, this system might have persisted. If, however, as Lack and Brown imply, the males were not only providing the females with no direct or indirect parental benefits, but were causing deleterious effects such as predator attraction, then females would be favored which tended to move away from the breeding ground to nest. In this hypothesis the traditional nature of the breeding grounds and considerable differential mating success among males could both precede the lek system as such. In a similar hypothesis females may have evolved early to remove themselves from their nesting ground so effectively with the onset of the breeding season that it became unprofitable for males to seek sexually responsive females there.

While these explanations may account for the separation of mating and nesting areas, they fail to account for aggregations that seem to serve no function for either sex other than mating. To solve this problem it may be profitable to consider certain general questions related to sexual selection (see also Otte, in press; Borgia, unpublished).

PARENTAL INVESTMENT, RESOURCE CONTROL, AND SEXUAL SELECTION

Among acoustical insects the only species in which the males are likely to be correctly viewed as investing parentally are the following:

1. some burrowing crickets in which the male gives up his burrow to a female after mating with her (West and Alexander 1963), and a South African species, whose males may even sometimes allow females with which they have mated to consume their bodies (Alexander and Otte 1967*a*);
2. some katydids in which the male's spermatophore is huge, as much as one-fourth his body weight (Busnel, Dumortier, and Busnel 1956), consisting chiefly of a gelatinous mass consumed by the female and perhaps properly viewed as a protein meal contributing to successful egg production and oviposition (see also Thornhill 1974).

When males do not invest parentally, one parameter of mate selection by females is removed from significance, and because polygyny is a correlate the reproductive variance of the males is likely to be greater than that of the females (Bateman 1948; Trivers 1972). This increased reproductive variance is important to the females, since it may be reflected in the reproductive success of their sons. Females in polygynous species in which the males do not invest parentally, including nearly all acoustical insects and fireflies, will thus evolve to judge males by their prowess in activities related to high mating success; this kind of selection will in turn reinforce trends toward intense mating competition among males.

Males of polygynous species may gain matings by controlling resources valuable to females (Sherman, unpublished) or by controlling areas where females are dense (Le Boeuf 1974). If males are unable to gain matings in either of these fashions, then another set of parameters by which females might usefully judge males has been eliminated. Males in such species may evolve to become either better searchers or better signalers. Searching seems the more likely strategy when females are relatively sedentary, as in web-building spiders. Evolution of signaling by males also seems unlikely in polygynous species if the males cannot control resources valuable to females or areas where females are dense. The female who mates with a male that has told her only that he is a sexually responsive male of her species has not been very selective.

Females in species whose males do not invest parentally and cannot profitably control resources valuable to the female or areas in which females are dense must judge males on their ability to locate or attract females and to mate successfully with such females. Female criteria for mate selection might include a male's vigor and searching ability, the intensity, persistence, or excellence of his signal, or the presence of other females in his vicinity.

Females in such species are also likely to gain by encountering the largest number of sexually active males in the shortest time with the least risk, thereby having the greatest opportunity to compare males and mate with a highly successful individual. The importance of the opportunity for the female to select among males may lead to tendencies by females to become more responsive in the presence of numerous males, to refuse to mate with lone males, or to seek out regions in which males are most dense. As a result of such behavior by females, males in aggregations or groups may do all or most of the mating. Larger aggregations or aggregations attracting females from greater distances may secure more females per male than others (see also Otte, in press), providing a reasonable selective background for Lack's (1968) second hypothesized function of leks (above).

The importance of the male signals in this model suggests that chorusing synchrony, which in *M. cassini* increases a group's attractiveness to females (Alexander and Moore 1958), is advantageous to each male because it increases the number of females attracted to his group. It implies that in some cases aggregation and synchrony may have evolved together in resource-based as well as non-resource-based male aggregations. Furthermore, it is at least possible that, in both acoustical insects and lek-breeding birds and mammals, female sexual selection may be powerful enough to cause breeding aggregations even though in such groups each male's and each female's susceptibility to predation is increased.

The breeding aggregations of periodical cicadas, honeybees, and the lek-breeding vertebrates mentioned above are likely to have been forced by female sexual selection. In birds and mammals the same locations may be remembered from year to year by both sexes; in honeybees certain kinds of terrain may be used by successive generations; in *Drosophila* certain locations on plants may be used. In acoustical insects the locations of choruses within habitats are less predictable, with areas of greatest emergence and relative predator invulnerability perhaps more important than other factors.

The term *lek* is often applied to the places where the animals in question aggregate, rather than to the aggregations themselves or their activity (e.g., Wiley 1973; Spieth 1974; Buechner and Roth 1974). But the significant generality appears to be the kind of aggregation rather than the specific location. If leks were localities, the term could not be applied easily to most insect mating assemblies. *Lek*, however, is derived from Swedish and implies play (*leka*), activity (-*lek*), or copulation (*leker*).[1] Thomson (1964) and others confirm this usage (i.e., as a kind of activity) and Otte (in press) uses the term in this apparently appropriate fashion. The words *arena* and *court* may be retained to describe the locations of leks.

Lek, then, appropriately refers to mating aggregations or assemblies, regardless of their locations or permanence of location. Some leks may have no function other than mating for either sex; they may be designated as non-resource-based or purely mating leks. Although such aggregations have been suggested previously, for lek-breeding birds and mammals and by the term *mating assemblies* (for example, Downes 1958 concluded that dipteran swarms "serve to concentrate the population, otherwise widely dispersed . . ."), it appears that such designations were casual results of our ignorance of the difficulties in explaining the adaptive significance of aggregations of males through benefits to individuals, rather than evidence of perception of the unusual selective background of strictly mating aggregations.

Leks with no function other than mating will be difficult to identify. Their failure to be associated with traditional locations in insects has probably concealed them in that group, where they should be prevalent owing to lack of parental investment by males. The possibility always exists that gain can be realized from tendencies to concentrate leks on food or oviposition sites, or in locales that are

[1]Ann E. Pace: personal communication.

relatively predator safe, even for individuals not chorusing or mating. Thus mating clusters of the cicada, *Diceroprocta olympusa*, in Florida resemble those of *Magicicada* species but characteristically occur on an aromatic shrub, Hercules' Club (*Xanthoxylum clava-herculis*) (Alexander and Moore, unpublished data). Dipteran swarms occur near oviposition sites, but "markers" resembling such sites may suffice (Downes 1969). Hence mating assemblies tend to shift in and out of the category of being purely products of the pressure by females for opportunities to compare males in the absence of male control of resources, and often it will be difficult to characterize mating aggregations in regard to these questions.

I suggest that resource-based mating aggregations, in which males cluster around resources valuable to females or in regions where females are dense for reasons other than mating alone, be included in the concept of *lek* whenever there is reasonable evidence that the males actively aggregate—that is, whenever males and groups of males can be shown to attract other males as well as females. As with purely mating leks, such aggregations are likely to involve much sexual selection by females, and to reflect a history of females favoring groups of males over single males and larger groups over smaller groups. Males in resource-based leks as well as those in non-resource-based leks are likely to gain from cooperating to increase female attendance at the lek.

EVIDENCE OF SEXUAL SELECTION IN CHORUSES

To human observers, male insects in choruses appear quite similar to one another, and, although they are more difficult to observe individually than birds and mammals, there seems to be no evidence of the incredible mating differentials that occur in vertebrate leks (e.g., Wiley 1973). Vertebrate males differ obviously from insect males in several ways that could account for their greater differentials in mating success.

First, lekking vertebrates are iteroparous across two or more breeding seasons, while for chorusing insects the breeding season is short and continuous, extending across a few weeks or months, and often essentially coincident with the period of adult life. This means that individual recognition is less likely to be important in insects and that male insects are more restricted in ability to progress gradually toward the best location in a lek as they grow, learn, and increase in dominance status.

The long adult life and iteroparity of vertebrates also provide multiple opportunities for the favoring of alleles improving the ability of their possessors to utilize innumerable environmental variables to adjust their phenotypes advantageously in the race of sexual selection.

Finally, young vertebrates receive considerable parental care from at least the female, as compared with chorusing insects. This parental care is an additional source of phenotypic differences between males (hence their differential reproduction as a result of selection by females) even after genetic uniformity among males is approached as a result of extreme polygyny. Some differences in male success may depend chiefly upon differences in the maternal behavior of their mothers, including differential investment by the mother in offspring of the two sexes (Trivers and Willard 1973).

In dipteran swarms, mating typically requires but a few seconds, and a male scarcely needs to drop out of the swarm to mate (Downes 1969). Cicada matings, on the other hand, sometimes occupy two or three hours, and the process of pair formation may take equally long (Alexander 1968). This particular feature of cicada choruses suggests that more selection of mates by females may be going on in the chorus than is evident. If so, we can only speculate about what attributes vary among males, for variations are not obvious to the human observer, as they are in bird leks.

If the unusually long period of copulation is due to the advantage to the male in removing the female from the chorus as long as possible, then perhaps the female's coyness is related to the relatively great investment she makes with whatever male she mates, considering the likelihood that she will be forced to remain in copula with him for a long period and the fact that females mate more than once (Alexander 1968). Again, how she might identify a suitable or superior male is totally unknown. Alternatively, the long period of "courtship" may reflect the value to a male of remaining near any female, because of the probability that she will become sexually responsive before they are accidentally separated. Unfortunately, because we were not aware of these problems when most of the periodical cicada work was done, there are no data from which to test these or other alternatives. The question of sexual selection within mating aggregations of insects thus remains open.

IS KIN SELECTION INVOLVED IN CHORUSING AGGREGATIONS?

Wild turkeys are reported to travel in sibling groups during the breeding season, with most males defending a dominant brother while he mates (Watts 1969; Watts and Stokes 1971). Chorusing insects are unlikely to recognize their brothers and profit by kin selection in the fashion suggested by the turkey example. There is a sense, however, in which average relatedness of males could be involved in their chorusing tendencies. Both sexes may gain from outbreeding, but the lower intensity of sexual competition among females in polygynous species means that they are more likely to evolve special behavior promoting outbreeding, whereas males profit from mating with any available responsive female. If cicada populations are somewhat viscous, we may envision groups of males evolving to sing where they emerge, and cooperating as signaling groups to outcompete other groups of males emerging elsewhere. Even if females disperse to mate, a tendency to oviposit near where they mate could lead to males being more closely related to members of their natal group than to males elsewhere. If females gain more from outbreeding than do males, their dispersal to this end may promote the tendency of males to compete as groups to attract females moving between choruses.

CONCLUSIONS

With the exception of the recent work of a few authors, studies on insect acoustical behavior, as with studies on mating assemblies in general, have tended to ascribe function at the population level rather than the individual level, and thus have failed to discern evidence of male-male competition and to elucidate functions in sexual selection.

Synchrony, alternation, and other phonoresponses which sometimes occur in insect choruses appear to be mostly explainable as competitive interactions between neighboring males. Insect choruses are probably most often incidental effects deriving from the aggregate of such responses when large numbers of males are for one reason or another closely spaced. Most mating "assemblies," including insect acoustical choruses, may ultimately be explained as males competing for females at oviposition, hibernating, feeding, or other locations where females are likely to occur.

Among choruses of acoustical insects, those most difficult to explain in this fashion are produced by the periodical cicadas. Males are known to be attracted to choruses in large numbers, and many if not most females appear to be attracted from outside the choruses; there is no obvious control by males of resources valuable to females, and no obvious tendency to form choruses where females are most dense. A model is proposed which suggests that males in these species have evolved to compete by cooperating in groups, including chorusing in synchrony, because of the importance to the females of comparing males at mating time, and resulting tendencies in females to mate only where males are dense. Such female selection is expected to force male aggregation even in the absence of ability by males to control resources valuable to females or areas where females are likely to be dense, and even, sometimes, when individual males and females both are thereby rendered more susceptible to predation. Once mating is largely or entirely restricted to male aggregations, either resource-based or non-resource-based (i.e., those in which mating is the sole function for both sexes), every male profits from cooperation, such as synchrony in chorus, which increases the number of females attracted to his particular group.

In insects, mating aggregations cannot be restricted to traditional (remembered) sites, as they appear to be in some vertebrates, and there is as yet no evidence of the enormous differences in male reproductive success reported in bird and mammal leks. In both vertebrates and invertebrates, however, the *lek* concept is appropriately applied to actively aggregating male groups within which all or nearly all mating occurs.

ACKNOWLEDGMENTS

I am deeply indebted to the following people who have worked on this general topic during the past several years: Gerald Borgia, Marianne N. Feaver, Richard D. Howard, James E. Lloyd, Glen K. Morris, Daniel Otte, Ann E. Pace, and Thomas J. Walker. All of these individuals gave freely of unpublished ideas and data, and all allowed me access to unpublished manuscripts. Borgia, Feaver, Howard, Otte, and Lloyd all criticized an early draft of the manuscript. William Cade allowed me to cite an unpublished observation. I am particularly indebted to Gerald Borgia, with whose ideas and arguments parts of the last portion of the paper became remarkably convergent. Ann Pace provided references on anuran behavior, and assisted with the derivation of the term *lek* by communicating directly with Cheri Register of the Scandinavian Department of the University of Minnesota.

REFERENCES

Alexander, R.D. 1956. A comparative study of sound production in insects, with special reference to the singing Orthoptera and Cicadidae of the eastern United States. Ph.D. dissertation, Ohio State University.

———. 1960. Sound communication in Orthoptera and Cicadidae, p. 38-92 and 5 recordings. *In* W.E. Lanyon and W.N. Tavolga [eds.], Animal sounds and communication. Amer. Inst. Biol. Sci. Publ. 7.

———. 1961. Aggressiveness, territoriality, and sexual behaviour in field crickets (Orthoptera: Gryllidae). Behaviour 17:130-223.

———. 1962. Evolutionary change in cricket acoustical communication. Evolution 16:443-67.

———. 1967. Acoustical communication in arthropods. Annu. Rev. Entomol. 12:495-526.

———. 1968. Arthropods, p. 167-216. *In* T.A. Sebeok [ed.], Animal communication. Univ. Indiana Press, Bloomington.

———. 1969. Comparative animal behavior and systematics, p. 494-517. *In* Systematic biology. Proc. Int. Conf., Nat. Acad. Sci. Publ. 1692.

———. 1974. The evolution of social behavior. Annu. Rev. Syst. Ecol. 5:325-83.

Alexander, R.D., and T.E. Moore. 1958. Studies on the acoustical behavior of seventeen-year cicadas (Homoptera: Cicadidae: *Magicicada*) Ohio J. Sci. 58:107-27.

———. 1962. The evolutionary relationships of 17-year and 13-year cicadas, and three new species (Homoptera, Cicadidae, *Magicicada*) Univ. Mich. Mus. Zool. Misc. Publ. 121:1-59.

Alexander, R.D., and D. Otte. 1967*a*. The evolution of genitalia and mating behavior in crickets (Gryllidae) and other Orthoptera. Univ. Mich. Zool. Misc. Publ. 133:1-62.

———. 1967*b*. Cannibalism during copulation in the brown bush cricket, *Hapithus agitator* (Gryllidae). Fla. Entomol. 50:79-87.

———. The crickets of Australia: systematics and biology. Unpublished ms.

Alexander, R.D., and E.S. Thomas. 1959. Systematic and

behavioral studies on the crickets of the *Neombius fasciatus* group (Orthoptera: Gryllidae: Nemobiinae) Ann. Entomol. Soc. Amer. 52:591-605.

Axtell, R.W. 1958. Female reaction to the male call in two anurans (Amphibia). Southwest. Natur. 3:70-76.

Bateman, A.J. 1948. Intra-sexual selection in *Drosophila*. Heredity 2:349-68.

Borgia, G. Resource structure and the evolution of mating systems. Unpublished ms.

Borror, D.J. 1954. Audio-spectrographic analysis of the song of the cone-headed grasshopper *Neoconocephalus ensiger* (Harris)(Orthoptera: Tettigoniidae) Ohio J. Sci. 54:297-303.

Brown, J.L. 1964. The evolution of diversity in avian territorial systems. Wilson Bull. 76:160-69.

Buck, J.B. 1935. Synchronous flashing of fireflies experimentally induced. Science 81:339-40.

Buck, J.B., and E. Buck. 1968. Mechanism of rhythmic synchronous flashing of fireflies. Science 159:1319-27.

Buechner, H.K., and H.D. Roth. 1974. The lek system in Uganda kob antelope. Amer. Zool. 14:145-62.

Busnel, R.G., B. Dumortier, and M.C. Busnel. 1956. Recherches sur le comportement acoustique des Éphippigères (Orthoptères, Tettigoniidae). Bull. Biol. Fr. Belg. 90: 219-86.

Busnel, R.G., and W. Loher. 1961. Déclenchement de phonorésponses chez *Chorthippus brunneus* (Thunb.) Acustica 11:65-70.

Busnel, R.G., F. Pasquinelly, and B. Dumortier. 1955. Le trémulation du corps et la transmission aux supports des vibrations en résultant comme moyen d'information a courte portée des Éphiggères mâles et femelles. Bull. Soc. Zool. Fr. 80:18-22.

Crook, J.H. 1965. The adaptive significance of avian social organization. Symp. Zool. Soc. London 14:181-218.

Downes, J.A. 1958. Assembly and mating in the biting Nematocera. Proc. Tenth Int. Congr. Entomol., Montreal 2:425-34.

———. 1969. The swarming and mating flight of Diptera. Annu. Rev. Entomol. 14:271-98.

Dumortier, B. 1963. Ethological and physiological study of sound emissions in Arthropoda, p. 583-654. *In* R.G. Busnel [ed.], Acoustic behaviour of animals. Elsevier, New York.

Eibl-Eibesfeldt, I. 1954. Paarungsbiologie der Anuren, Grasfrosch, Erdkröte, Laubfrosch, Wasserfrosch. 16-mm film. Institut für den Wissenschaftlichen Film, Göttingen, West Germany.

Emlen, S.T. 1968. Territoriality in the bullfrog, *Rana catesbeiana*. Copeia 1968:240-43.

Fisher, R.A. 1958. The genetical theory of natural selection. Dover, New York.

Gadgil, M. 1972. Male dimorphism as a consequence of sexual selection. Amer. Natur. 106:574-80.

Grove, D.G. 1959. The natural history of the angular-winged katydid *Microcentrum rhombifolium*. Ph.D. dissertation, Cornell Univ.

Hamilton, W.D. 1964. The genetical evolution of social behaviour. I, II. J. Theor. Biol. 7:1-52.

———. 1970. Selfish and spiteful behaviour in an evolutionary model. Nature 228:1218-20.

Hamilton, W.J. 1968. Social aspects of bird orientation mechanisms, p. 57-71. *In* R.M. Storm [ed.], Animal orientation and navigation. Proc. 27th Annu. Biol. Colloq. Oregon State Univ. Press, Corvallis.

Hill, G.K. 1974. Acoustic communication in the Australian field crickets *Teleogryllus commodus* and *T. oceanicus*. Ph.D. thesis, Univ. of Melbourne, Victoria, Australia.

Hogan-Warburg, A.J. 1966. Social behavior of the ruff, *Philomachus pugnax* (L.). Ardea 54:109-29.

Jones, M.D.R. 1966. The acoustic behaviour of the bush cricket *Pholidoptera griseoaptera*. I. Alternation, synchronism and rivalry between males. J. Exp. Biol. 45:15-30.

Lack, D. 1968. Ecological adaptations for breeding in birds. Methuen, London.

Le Boeuf, B.J. 1974. Male-male competition and reproductive success in elephant seals. Amer. Zool. 14:163-76.

Leuthold, W. 1966. Variations in territorial behavior of Uganda kob, *Adenota kob thomasi*. Behaviour 27:215-58.

Lin, N. 1963. Territorial behaviour in the cicada killer wasp, *Sphecius speciosus* (Drury)(Hymenoptera: Sphecidae). I. Behaviour 20:115-33.

Littlejohn, M.J., and A.A. Martin. 1969. Acoustic interaction between two species of leptodactylid frogs. Anim. Behav. 17:785-91.

Littlejohn, M.J., and G.F. Watson. 1974. Mating call discrimination and phonotaxis by females of the *Crinia laevis* complex (Anura: Leptodactylidae). Copeia 1974:171-75.

Lloyd, J.E. 1971. Bioluminescent communication in insects. Annu. Rev. Entomol. 16:97-122.

———. 1973*a*. Fireflies of Melanesia: bioluminescence, mating behavior, and synchronous flashing (Coleoptera: Lampyridae). Environ. Entomol. 2:991-1008.

———. 1973*b*. Model for the mating protocol of synchronously flashing fireflies. Nature 245:268-70.

———. Bioluminescence and communication. *In* T.A. Sebeok [ed.], How animals communicate. Univ. Indiana Press, Bloomington. In press.

Morris, G.K. 1970. Sound analyses of *Metrioptera sphagnorum* (Orthoptera: Tettigoniidae). Can. Entomol. 102:363-68.

———. 1971. Aggression in male conocephaline grasshoppers (Tettigoniidae). Anim. Behav. 19:132-37.

———. 1972. Phonotaxis of male meadow grasshoppers (Orthoptera: Tettigoniidae). J. N.Y. Entomol. Soc. 80:5-6.

Morris, G.K., G.E. Kerr, and D.R. Gwynne. Calling song function in the bog katydid, *Metrioptera sphagnorum* (F. Walker)(Orthoptera: Tettigoniidae): female phonotaxis to normal and altered song. Z. Tierpsychol., in press.

Muul, I. 1968. Behavioral and physiological influences on the distribution of the flying squirrel, *Glaucomys volans*. Univ. Mich. Mus. Zool. Misc. Publ. 134.

Oldham, R.S. 1974. Mate attraction by vocalization in members of the *Rana pipiens* complex. Copeia 1974:982-84.

Otte, D. 1972. Simple *versus* elaborate behavior in grasshoppers: an analysis of communication in the genus *Syrbula*. Behaviour 42:291-322.

———. 1974. Effects and functions in the evolution of signaling systems. Annu. Rev. Ecol. Syst. 5:385-417.

———. Communication in Orthoptera. *In* T.A. Sebeok [ed.], How animals communicate. Univ. Indiana Press, Bloomington. In press.

Otte, D., and A. Joern. Insect territoriality and its evolution: population studies of desert grasshoppers on creosote bushes. J. Anim. Ecol., in press.

Otte, D., and J. Loftus-Hills. Chorusing in *Syrbula*: cooperation, competition, interference or concealment (Gomphocerinae)? Unpublished ms.

Otte, D., and J. Smiley. Synchronous flashing in Texas fireflies (*Photinus*: Lampyridae): a consideration of interaction models. Unpublished ms.

Pace, A.E. 1974. Systematic and biological studies of the leopard frogs (*Rana pipiens* complex) of the United States. Univ. Mich. Mus. Zool. Misc. Publ. 148.

Parker, G.A. 1970. The reproductive behavior and the nature of sexual selection in *Scatophaga stercoraria* L. (Diptera: Scatophagidae). IV. Epigamic recognition and competition between males for the possession of females. Behaviour 37:8-39.

Patterson, R.L. 1952. The sage grouse in Wyoming. Sage Books, Denver.

Rosen, M., and R.E. Lemon. 1974. The vocal behavior of spring peepers, *Hyla crucifer*. Copeia 1974:940-50.

Shaw, K.C. 1968. An analysis of the phonoresponse of males of the true katydid, *Pterophylla camellifolia* (Fabricius) (Orthoptera: Tettigoniidae) Behaviour 31:203-60.
Sherman, P.W. The mating system and social behavior of the Asian honeyguide, *Indicator xanthanotus*. Unpublished ms.
Spieth, H.T. 1974. Courtship behavior in *Drosophila*. Annu. Rev. Entomol. 19:385-405.
Spooner, J.D. 1968. Pair-forming acoustic systems of phaneropterine katydids (Orthoptera: Tettigoniidae). Anim. Behav. 16:197-212.
Strang, G.E. 1970. A study of honey bee drone attraction in the mating response. J. Econ. Entomol. 63:641-45.
Thomson, A.L. 1964. A new dictionary of birds. Thomas Nelson & Sons, London.
Thornhill, A. 1974. Evolutionary ecology of Mecoptera (Insecta). Vols. 1, 2. Ph.D. thesis, Univ. Mich.
Trivers, R.L. 1971. The evolution of reciprocal altruism. Quart. Rev. Biol. 46:35-57.
———. 1972. Parental investment and sexual selection, p. 136-79. *In* B. Campbell [ed.], Sexual selection and the descent of man: 1871-1971. Aldine Publ. Co., Chicago.
———. 1974. Parent-offspring conflict. Amer. Zool. 14: 249-64.
Trivers, R.L., and D.E. Willard. 1973. Natural selection of parental ability to vary sex ratio of offspring. Science 179:90-92.
Walker, T.J. 1957. Specificity in the response of female tree crickets (Orthoptera, Gryllidae, Oecanthinae) to calling songs of the males. Ann. Entomol. Soc. Amer. 50:626-36.
———. 1969. Acoustic synchrony: two mechanisms in the snowy tree cricket. Science 166:891-94.
———. 1974. Character displacement and acoustical insects. Amer. Zool. 14:1137-50.
———. Speculations on the evolution of the acoustic behavior of Uhler's katydid. Unpublished ms.
Walker, T.J., and D. Dew. 1972. Wing movements of calling katydids: fiddling finesse. Science 178:174-76.
Watts, C.W. 1969. The social organization of wild turkeys on the Welder Wildlife Refuge, Texas. Ph.D. thesis, Utah State Univ.
Watts, C.W., and A.W. Stokes. 1971. The social order of turkeys. Sci. Amer. 224:112-18.
West, M.J. 1969. The social biology of polistine wasps. Univ. Mich. Mus. Zool. Misc. Publ. 140:1-101.
West, M.J., and R.D. Alexander. 1963. Sub-social behavior in a burrowing cricket *Anurogryllus muticus* (De Geer)

(Orthoptera: Gryllidae). Ohio J. Sci. 63:19-24.

West-Eberhard, M.J. The evolution of social behavior by kin selection. Quart. Rev. Biol., in press.

Wiley, R.H. 1973. Territoriality and non-random mating in sage grouse, *Centrocercus urophasianus*. Anim. Behav. Monogr. 6:87-169.

Williams, G.C. 1966*a*. Adaptation and natural selection. Princeton Univ. Press, Princeton.

———. 1966*b*. Natural selection, the costs of reproduction, and a refinement of Lack's principle. Amer. Natur. 100: 687-90.

Wynne-Edwards, V.C. 1962. Animal dispersion in relation to social behaviour. Oliver and Boyd, Edinburgh.

Zaretsky, M.D. 1972. Specificity of the calling song and short term changes in the phonotactic response by female crickets *Scapsipedus marginatus* (Gryllidae). J. Comp. Physiol. 79:153-72.

Zmarlicki, C., and R.A. Morse. 1963. Drone congregation areas. J. Apicult. Res. 2:64-66.

INSECT COMMUNICATION—CHEMICAL

Wendell L. Roelofs

People usually react in a threatened way when the terms *insects* and *chemicals* are used together. The prevailing impression seems to be that chemicals are poisons produced by chemists and spewed out by entomologists to kill insects and pollute the environment. More generally, chemicals are thought of as artificial creations added to our food and water as threats to our health. Many people seem to have lost sight of the fact that we and the world around us are composed of masses of chemicals. Chemicals are the constituents of the air, the water, our food and our bodies. Additionally, chemical messengers are involved in all our actions, thoughts, and bodily processes, from internal chemical signals that—even in the embryonic stage—regulate the metabolism of cells and the differential expression of its genes, to external chemical odors that massage our emotions in ways so subtle that we are not even aware of the stimuli.

Insects, too, are just masses of chemicals under the direction of chemical messengers, both internal and external. New vistas of research have been opened up on some of the insect messenger systems, but many areas are yet unexplored. I will concentrate in this paper on external chemical signals. The chemicals, called pheromones, are signals that operate between individuals of the same species.

Naturalists have been aware for many years that there is some kind of distance attraction between the sexes of some insects. In the early 1800s von Siebold (1837) described some female glands that were thought to emit odors to attract males. Evidence then was obtained by the great French naturalist, J. Henri Fabre (translation in Fabre 1916), that long-distance attraction of male moths to a caged female giant silkmoth was due to female effluvia, not to sight or sound. He wrote a very interesting account of some experiments he conducted in which his house became filled with the huge fluttering males attracted from miles around. By the turn of the century, a number of scientists had reported on attractant gland structures and made observations on the attraction phenomenon.

Speculation on the possibilities of utilizing this vulnerable part of the mating system for insect control started to grow, but not much effort was directed to this field of research until two notable reports came out in 1959 and 1962. In the first, Professor Butenandt and coworkers (1959) in Germany culminated thirty years of laborious research on the silkworm moth, *Bombyx mori*, with the identification of the first sex-pheromone structure. They accomplished this task without the aid of the highly sensitive instrumentation that was just coming into general use by chemists. The second important report was the book *Silent Spring*, by Rachel Carson (1962). In presenting vivid accounts of the dangers of misusing pesticides, it helped in directing the attention of scientists and federal funding agencies to alternative methods of insect control.

The identification of one sex-pheromone structure and the availability of new instrumentation and research funds opened the door on this area of research, and a herd of researchers jammed the doorway. Some have moved on through and have contributed excellent research in this field. There have been many recent reviews on this subject (e.g., Birch 1974; Jacobson 1974; Evans and Green 1973; MacConnell and Silverstein 1973; Priesner 1973; Shorey 1973; Jacobson 1972; Law and Regnier 1971); I will not give another such review here.

I will touch on some of the vistas that have opened up in pheromone research and discuss their relevance to entomologists. I will use our research on the redbanded leafroller moth, *Argyrotaenia velutinana*, to illustrate the various areas, without implying that it is always representative of the more extensive research conducted by others on other species of insects. The redbanded leafroller moth is a small moth in the family Tortricidae. Interestingly, it became a major pest of apple in eastern United States only after DDT was extensively used for codling moth control. DDT was ineffective against the redbanded leafroller moth, but it was toxic to the species' natural enemies. Since then the redbanded leafroller moth has become the most important tortricine pest of apple in eastern United States (Chapman and Lienk 1971).

PHEROMONE EMISSION

The Pheromone Gland

Research on the chemical communication system of this

species has been conducted with both academic and applied interests in mind. First, let us consider the role played by the female. It is easy to determine that the females assume the "calling" stance and protrude a gland at the abdominal tip that releases the sex pheromone. But more can be learned about the structure and secretory nature of the pheromone gland by using light, electron, and scanning electron microscopy.

Development of Pheromone Gland in the Pupal Stage. We were interested to find that the pheromone gland of the red-banded leafroller developed from the last four larval segments (Feng 1972). A tanned female pupa, four hours after the larval-pupal ecdysis, shows externally six distinct abdominal segments plus one cone-shaped terminal segment resulting from a fusion of the last four larval segments (fig. 1). The epidermis in a 24-hour-old pupa is still attached to the pupal cuticle, and there is no significant segmentation observable morphologically and histologically posterior to the sixth abdominal segment. Removal of the pupal cuticle from a 42-hour-old pupa reveals the cone-shaped segment as a tapered fingerlike projection. Two grooves demarcating the seventh and eighth segments appear on the pupal caudal end due to the retraction of the abdominal epidermal tissue. After 52 hours the grooves are deepened and the ovipositor begins to develop. Sixty hours after the larval-pupal moult, the fingerlike projection is retracted beneath the eighth segment and becomes flattened tissue between the eighth segment and the enlarged ovipositor lobes. Differentiation of the pheromone gland begins in the 72-hour-old pupa on the dorsal intersegmental epithelium between the eighth and terminal segments. Throughout fourth and fifth days of the pupal stage, the ovipositor lobes become flatter and sclerotized, the pheromone gland invaginates beneath the eighth segment, and the caudal end continues to telescope into the body cavity. The invagination appears as a layer of simple cuboidal epithelium with a thin cuticle laid on the apical surface.

Adult Sex Pheromone Gland. The pheromone gland reaches its maximum size two to three days after moth emergence. Abdominal heomolymph pressure extrudes the gland as a white membranous sac (fig. 2A). The gland extends into the body cavity directly beneath the eighth tergum when not extended, and can be seen as a transparent pinkish, convoluted sac, full of large dark-red staining nuclei, after a Feulgen reaction (fig. 2B). A histological section of the fully developed pheromone gland reveals it as a single highly

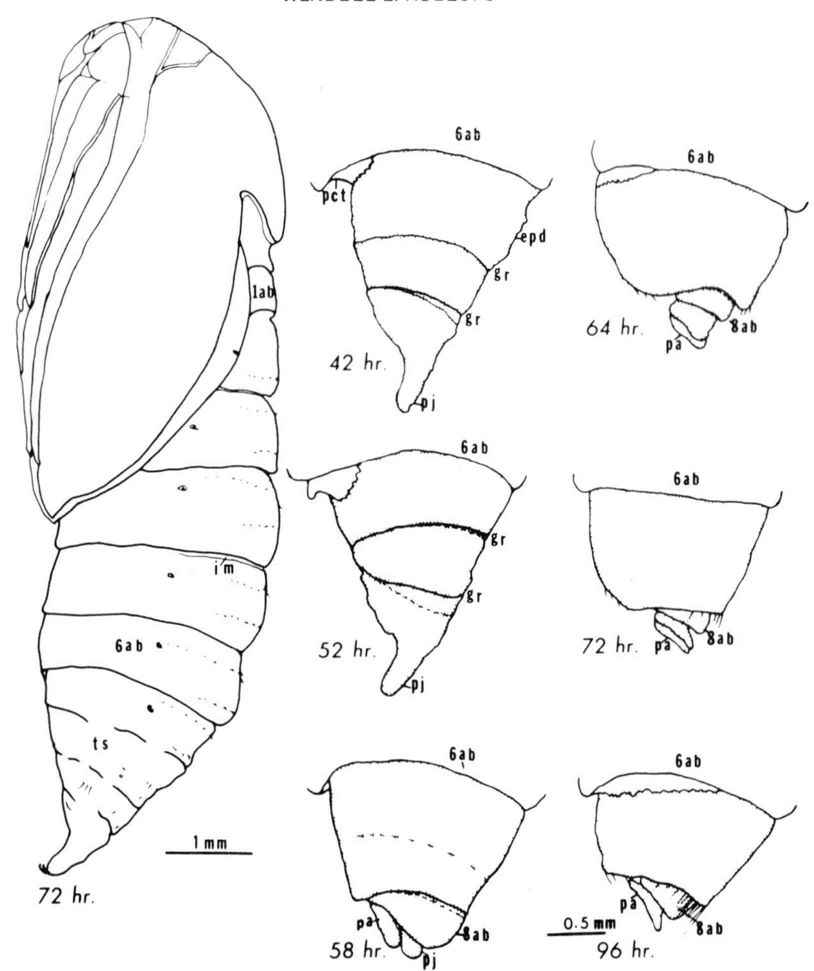

Fig. 1. Development of sex pheromone gland in A. velutinana *pupal stage. Pupal cuticle has been removed from terminal segments after sixth abdominal segment (*6ab*). The grooves (*gr*) are seen in the 42-hour pupa, and the caudal projection (*pj*) is then shown to withdraw gradually between the eighth segment and the newly formed padlike papillae anales (*pa*) (Feng 1972).*

convoluted layer of greatly enlarged epidermal cells which are continuous with the epidermis of the intersegmental membrane between the eighth and terminal abdominal segments (fig. 2C). The glandular epithelium is composed of tall,

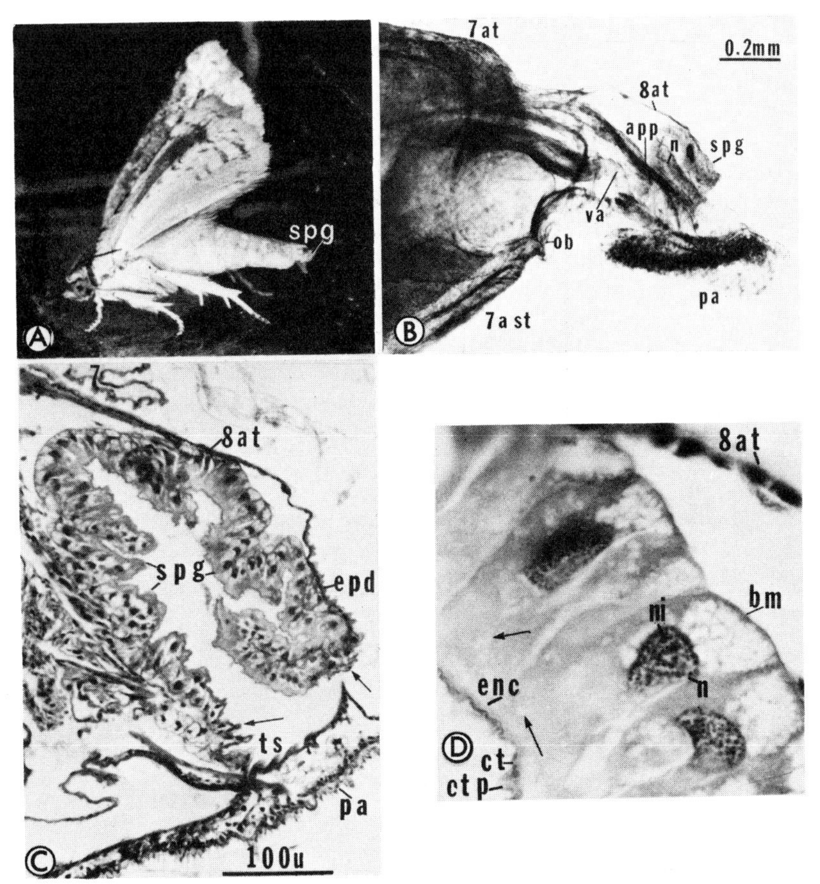

Fig. 2. Female sex pheromone gland (spg) of A. velutinana A. Female in the calling position with wings lifted and papillae anales (pa) bent downwards to evert the gland. B. Whole mount with Feulgen preparation showing the pheromone gland as a convoluted sac with large prominent nuclei (n). C. Sagittal section of caudal end showing the gland as a convoluted, enlarged epithelium between the eighth (8at) and terminal abdominal segments (ts). D. Pheromone gland showing tall columnar cells with nuclei and nucleoli (ni); the basal cytoplasm appears vacuolated and the apical surface is covered with a layer of cuticle (ct) and some cuticular processes (ctp) (Feng 1972).

columnar cells (30-44 μ) with the basal cytoplasm appearing to be highly vacuolated (fig. 2D). Electron microscopy also reveals an abundance of vesicles (1-6 μ) in the basal half of the cell. They appear to be lipid droplets because of their smooth, electron-lucid appearance and location, and could be the accumulation of pheromone chemicals and their precursors. Large increases of endoplasmic reticulum in adult sex-pheromone glandular cells suggest that the endoplasmic reticulum system is the site of production of the attractant compound.

An investigation by scanning electron microscopy of the pheromone-gland surface shows that it possesses a tremendous amount of surface area for pheromone evaporation. The surface of the highly irregular pheromone gland is different from the other intersegmental membrane (fig. 3), and when

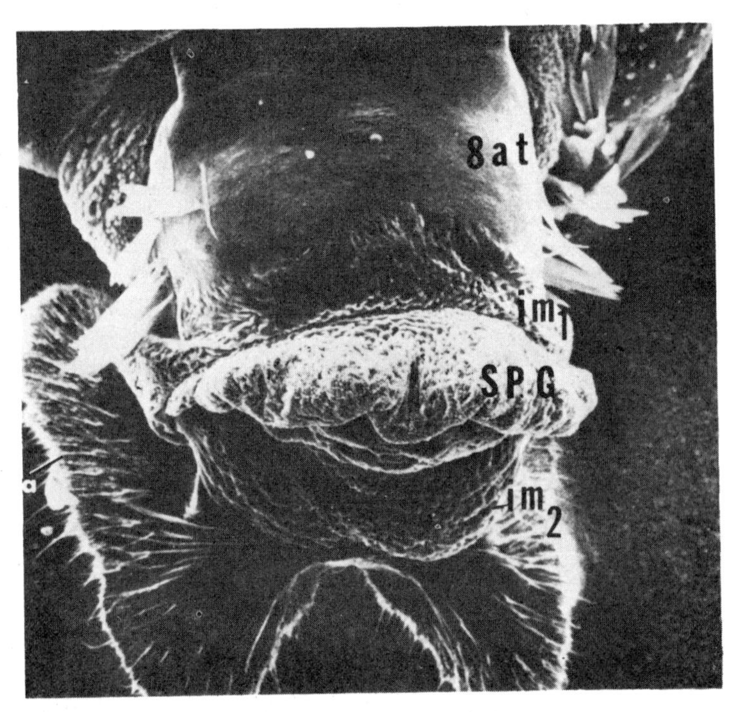

Fig. 3. Scanning electron micrograph of the caudal end of an A. velutinana *female showing the pheromone gland as different from the other intersegmental membranes* (im_1 *and* im_2) *(Feng 1972).*

extruded it exhibits a convoluted surface with a definite polygonal corolla pattern (fig. 4). Each polygonal feature

Fig. 4. Scanning electron micrograph (1510x) of the pheromone gland of A. velutinana *showing the polygonal corolla pattern* (pg)*(Feng 1972).*

appears to coincide with an underlying epidermal cell. High magnification reveals additional surface area from intra- and intercellular ridges. These studies could serve as the basis for further investigations on the actual site of pheromone production and its transport from the cell, on the biosynthetic pathways utilized to produce the pheromone components (including research on the possibility of a rhythm of synthesis corresponding to female calling or on the production of inactive precursors that are stored and converted to active compound when released from the cell at the time of calling), and on the controlling mechanisms of enzyme systems responsible for producing particular mixtures of various isomers and analogs. But I feel that we now have pushed this particular area to the limits of most entomologists, so let us return to the insect.

Control of Female Calling

An interesting area of research is available to entomologists in describing what factors control and affect the active release of the pheromone from the female moth (reviewed by Shorey 1974). In general, pheromone emission from female moths appears to be on a circadian rhythm entrained by the daily light-dark cycle. The tendency for females to call is also increased within certain optimum temperature and air-velocity ranges. However, temperature changes within this range can modify the timing of the circadian rhythm. For example, in the laboratory (Cardé et al. 1975) the redbanded leafroller females initiate calling one hour after darkness at 24° under a 16:8 LD (fig. 5). At 16° the females call much earlier (about six hours

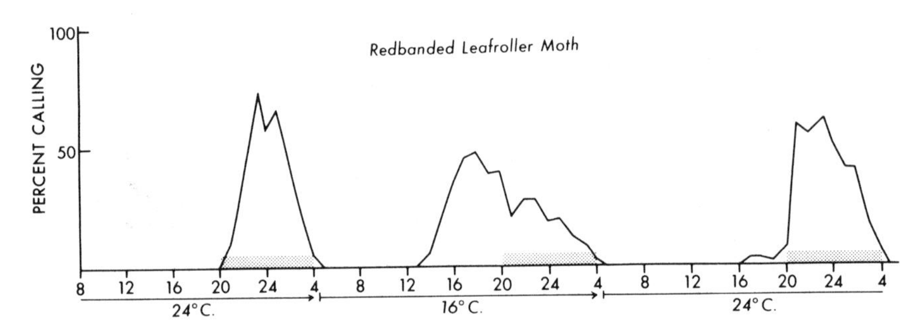

Fig. 5. Effect of temperature on "calling" with female A. velutinana *under a 16:8 LD.*

prior to scotophase). This phenomenon is also observed with this insect under natural conditions. In the cool weather of the spring flight, attraction occurs in the afternoon, whereas attraction occurs after sunset during the July summer flight.

Research by a number of investigators has shown that the influence of the various environmental and physiological factors that affect female calling varies from species to species. It is obvious that much more research is needed by entomologists to understand better what causes female moths to protrude their pheromone gland with hopes of luring male moths for mating.

Pheromone Chemistry

There is another area of research in which the female moths are used rather than studied. This involves the structural identification of the pheromone components. The components can be obtained from excised female pheromone glands, or from airborne collections of calling-female effluvia.

Pheromone Components of the Redbanded Leafroller Moth. After two years of research and about fifty thousand female tips, we identified the first component from the redbanded leafroller moth to be *cis*-11-tetradecenyl acetate (Roelofs and Arn 1968). Field screening of various compounds showed that dodecyl acetate could be used to increase greatly the number of attracted males (Roelofs and Comeau 1968, 1971*b*). Subsequently, field testing of isomers in Iowa by Dr. Klun and coworkers (1973) showed that about 7% *trans*-11-tetradecenyl acetate was also needed for attractancy. We have recently identified all three pheromone components (fig. 6) from female pheromone gland extract and effluvia and have defined the optimum ratios for field attractancy (Roelofs, Hill, and Cardé, in press).

REDBANDED LEAFROLLER MOTH

(Argyrotaenia velutinana)

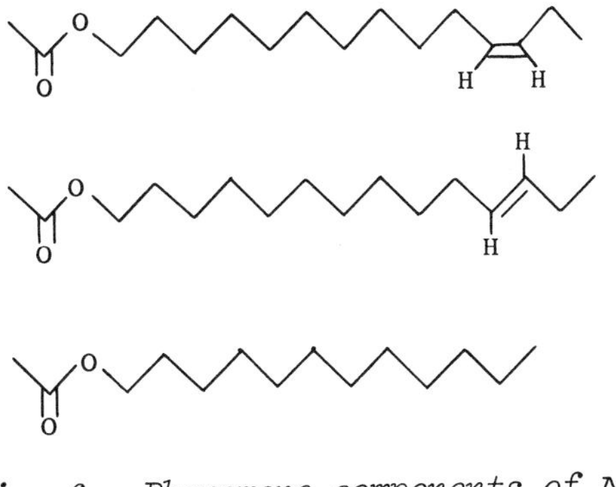

cis-11-14:Ac

trans-11-14:Ac (8%)

12:Ac

Fig. 6. Pheromone components of A. velutinana.

Pheromones of Tortricid Species. The chemical studies can give some insight into questions of pheromone evolution and species reproductive isolation. For example, by 1971 we had accumulated data on the main attractant chemical for a number of species in the two tortricid subfamilies, Olethreutinae and Tortricinae (Roelofs and Comeau 1971*a*). It was interesting to find that the attractants of the former subfamily were all 12-carbon-chain compounds, whereas the attractants of the latter were all 14-carbon-chain compounds. As many of the tortricine species use the same compound, the role of pheromones in the reproductive isolation of these species was questioned. Recently, however, we have found that these species use precise blends of a number of components—some using various mixtures of *cis-trans* isomers, some utilizing acetate-alcohol or acetate-aldehyde mixtures, and some using mixtures of positional isomers (fig. 7).

The actual chemistry involved in this particular study is not too complicated, but when one considers the complexity of structures and the sophistication needed to identify and synthesize optically active enantiomers, as reported by Dr. Silverstein and coworkers (Plummer and Silverstein 1974), it becomes apparent that again we have moved out of the realm of entomology and into the realm of chemistry. There must always be a close interaction among the scientists, however, with advances being made together. The identification of the pheromones makes quantitative research in a multitude of areas more feasible.

PHEROMONE RECEIVER

Male Responsiveness

Research can be conducted with the males to determine which environmental and physiological variables control and affect their pheromone behavior. Research to date indicates that the times of female calling and of maximum male responsiveness are synchronized throughout the range of variables that affect both (Shorey 1974). For example (fig. 8), redbanded leafroller males (Cardé et al. 1975) will respond to pheromone about six hours prior to scotophase at 24°—similar to the temperature shift observed with calling females. An area of interesting research awaits the entomologists willing to carry out studies on the behavioral sequences exhibited by insects responding to their sex pheromone.

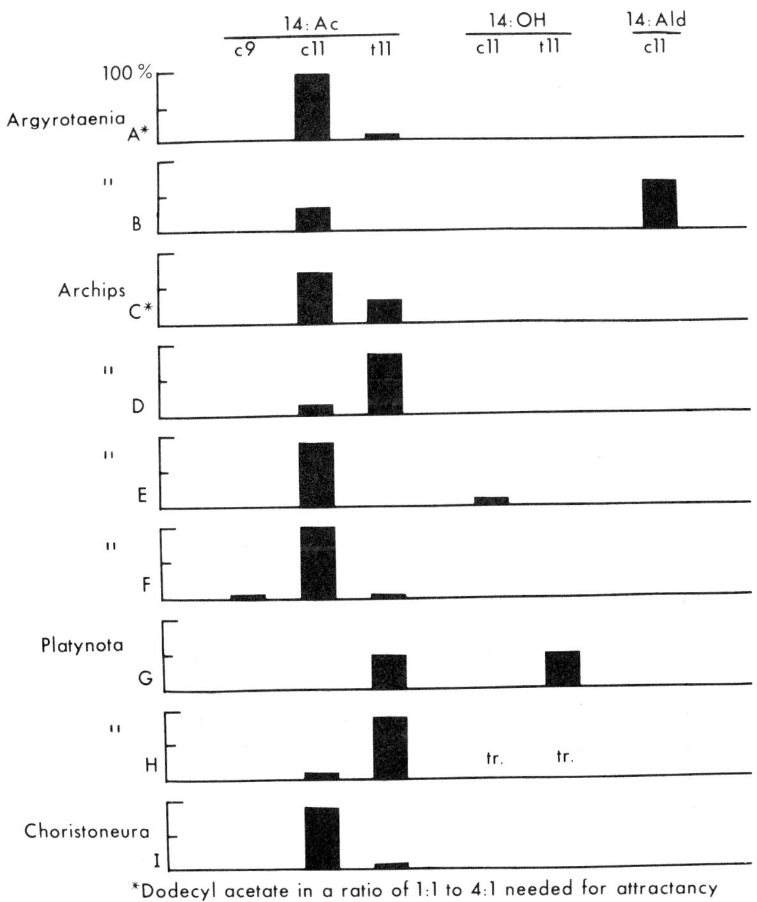

Fig. 7. Pheromone blends defined for various tortricid species in the Tortricinae subfamily.

Antennal Responses

The actual perception of the pheromone molecules by the male antennae involves another area of research (Payne 1974; Schneider 1974).

Electroantennograms. One can begin with studies of the whole antennal responses—a technique called the electroantennogram (EAG). This technique, developed primarily by Professor Schneider and coworkers in the 1950s, gives a measurement of the summated receptor potentials of a number

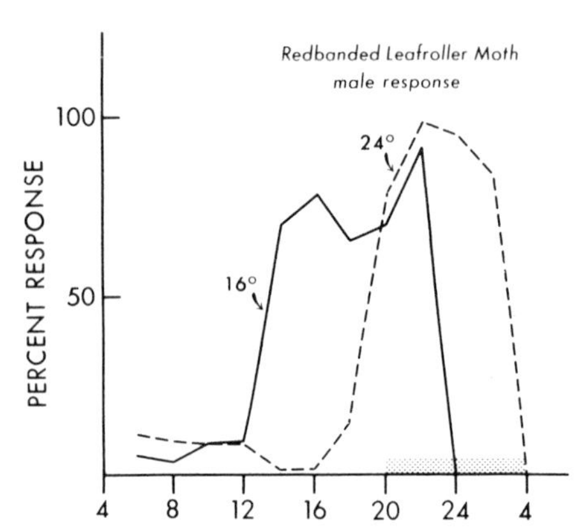

Fig. 8. Effect of temperature on the response of male A. velutinana *to pheromone.*

of olfactory receptors responding to a stimulus. A recording electrode is placed over the distal end of an antenna, and an indifferent electrode makes contact with the proximal end. The signals from the antenna are amplified and displayed on an oscilloscope.

It has been found that the majority of receptors in male moths are responsive to the sex pheromone. This allows us to use the antenna as a very sensitive instrument in defining pheromone components (Roelofs and Comeau 1971*c*). It can be used as a sensitive assay for fractions collected from the gas chromatograph (fig. 9).

In addition to locating the retention time of the primary pheromone component—the one usually located by laboratory behavioral assays—the EAG can also detect secondary pheromone components when mixtures are involved. Also, the primary pheromone component usually elicits the greatest antennal response in a series of related compounds. This has allowed us to use the profile of the antennal response to our library of hundreds of standards to predict the structure of many primary pheromone components, including information on functional groups, carbon-chain length, and the position and configuration of any double bonds involved. For example, a comparison of the antennal responses (fig. 10) from males of two *Argyrotaenia* species shows that *A. velutinana* (redbanded leafroller) responds primarily to *cis*-11-tetradecenyl acetate (the primary pheromone component), whereas *A. citrana* responds well to both *cis*-11-tetradecenyl acetate and *cis*-11-

tetradecenal. Further analysis and field tests proved these two compounds to be the pheromone components for this species (Hill et al., in press). With species utilizing a doubly unsaturated pheromone, the monounsaturated standards many times give an antennal response profile with two standards producing good responses (Roelofs et al. 1971, 1973). The two areas of activity are usually indicative of the position and configuration of both double bonds of the pheromone (fig. 11).

Single Cell Studies. To obtain more information on the antennal receptors, some investigators have initiated studies on the response of single cells. In research with the silkmoth and the gypsy moth, Professor Schneider and coworkers have described the pheromone-perceiving sense cells as odor specialists (see Kaissling 1971 and references therein). The acceptors are very specific for the pheromone molecule because of the high selective binding capacity of the acceptors. Chemicals with the correct arrangement of active sites would interact with the acceptors and trigger nerve impulses

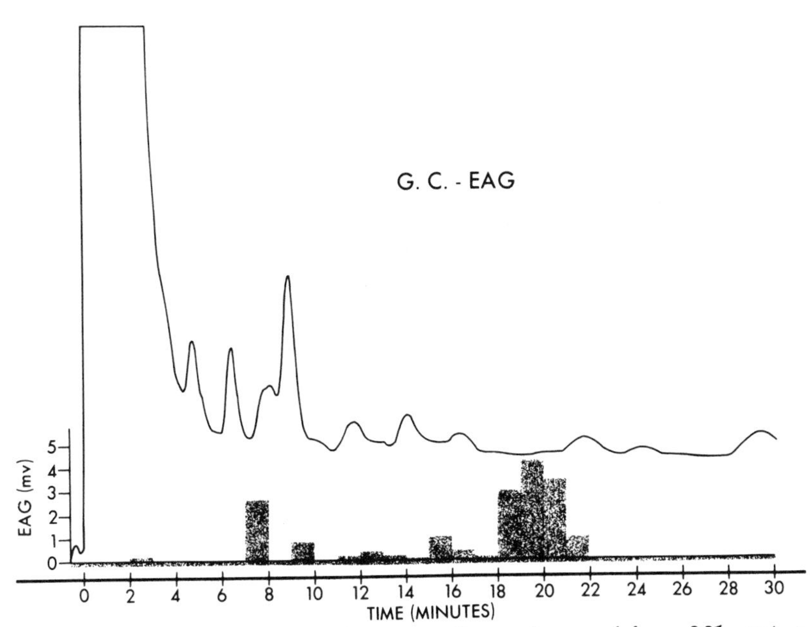

Fig. 9. Assaying by EAG the gas chromatographic effluent of female pheromone gland extract. Antennal responses indicate retention times of possible primary and secondary pheromone components.

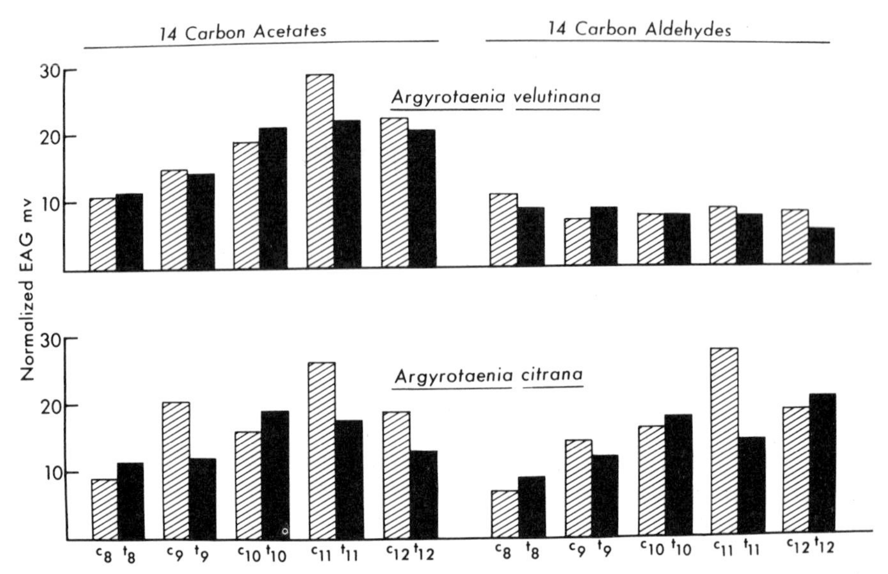

Fig. 10. Male antennal responses of two Argyrotaenia *species to monounsaturated 14-carbon acetate and aldehyde standards, showing* A. velutinana *responding best to* cis-*11-tetradecenyl acetate and* A. citrana *responding well to this* cis-*11 acetate and to* cis-*11-tetradecenal.*

in the receptor cell. The nerve impulses from these odor specialists would be sent to the brain with exact information on the presence and abundance of pheromone molecules in the air stream.

With the redbanded leafroller moth—an insect employing a pheromone blend instead of just one compound—Dr. O'Connell (1972, and unpublished) has obtained evidence that the sense cells are probably not specialists in terms of giving only one type of activation response. He found that the sensilla trichodea usually possess two olfactory receptors that are differentiated in the recording by a difference in spike amplitudes. One of the receptors (cell one, high amplitude) is usually more sensitive to the primary pheromone component (c11-14:Ac), whereas the other receptor (cell two, small amplitude) is usually more sensitive to the secondary component (t11-14:Ac). The latter compound mostly produced inhibitory responses in cell one, whereas the other secondary component (12:Ac) elicited no response by itself but increased the response frequencies when combined with c11-14:Ac. Further investigation with eight other compounds found to affect male attractancy in field studies showed that there was much qualitative and quantitative variation in response frequency from receptor to receptor. It also became apparent that the simple presence or absence of a certain amount of receptor activity could not account for the intrinsic activity of a pheromone molecule. The insect can apparently differentiate the compounds even though various quantities of different chemicals can produce the same absolute discharge magnitude in a receptor. O'Connell reports that the encoding of odor quality for the redbanded leafroller moth involves the pattern of activity across an ensemble of receptors with varying sensitivities.

Much more research needs to be done on pheromone perception and transduction mechanisms, but now we have crossed over into areas of neurophysiology in which most entomologists do not venture.

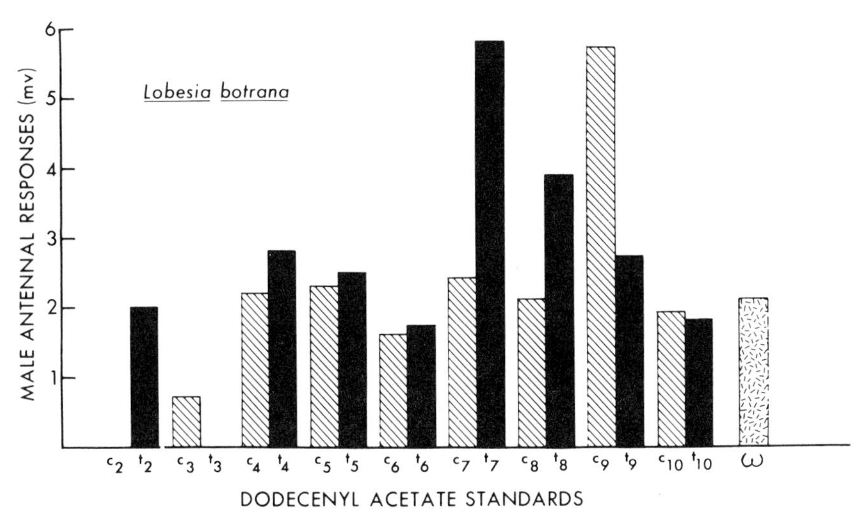

Fig. 11. Male Lobesia botrana *antennal responses to a series of monounsaturated 12-carbon acetates, showing good responses to* trans-*7-dodecenyl acetate and to* cis-*9-dodecenyl acetate. The pheromone is* trans-*7,*cis-*9-dodecadienyl acetate.*

PHEROMONES IN NATURE

Trap Studies

Let us consider some of the areas opened up in utilizing pheromones in field programs. The availability of potent pheromone traps has allowed entomologists to conduct research on monitoring the presence and abundance of particular insect populations. Information can be obtained on the type of trap, placement of traps, type of pheromone carrier, quantity and ratios of pheromone components, longevity of attractiveness, relationship of moth flight to larval hatch, relationship of moth catch to subsequent crop damage, and so on. This information can then be used by growers to time insecticide applications more precisely (Tette 1974).

Mass Trapping for Control

Efforts have been made to use pheromone traps as insect control devices (Birch 1974, chap. 22; Roelofs, in press). For example, throughout four years of mass trapping for redbanded leafroller males in 50 acres of apple (40 traps/acre), we were able to maintain this pest at commercially acceptable levels (Trammel, Roelofs, and Glass 1974). However, the complex of insect species requiring suppression in New York apple orchards precludes mass trapping as an economical method of control. There are a number of situations being investigated by others in which mass trapping may be feasible.

Mating Disruption for Control

Another technique which has gained a lot of support is the use of pheromones in mating disruption programs. The principle is to permeate the air with pheromone or inhibitor, or both, to disorient or habituate the male moths. Shorey and coworkers have pioneered this area, using widely separated pheromone evaporators for the cabbage looper and pink bollworm moths, and Beroza has coordinated efforts on the gypsy moth using microencapsulated pheromone formulations (Birch 1974, chap. 22). We have initiated research using both systems on the redbanded leafroller moth, and have obtained encouraging results from both. The use of microencapsulated formulations is the most appealing to us since the formulations can be applied with conventional spray equipment. Perhaps spraying five grams of encapsulated pheromone per acre every two or three weeks would effect mating disruption. The intriguing thing about this approach in terms of our leafroller complex in New York is that it may be possible to use one mixture of two or three chemicals to disrupt males in the whole complex. Unfortunately, data are not available on which pheromone component or components are the most effective for each species. It is not known if the proper pheromone blend must be used or if just one major or one minor component can be as effective.

In asking questions relating to mating disruption, one soon realizes that we do not know much about the role of the individual pheromone components. They are all needed for maximum trapping results, but are they all involved in long-distance attraction? Some components may operate over long distances and elicit upwind orientation, while other components may only be effective later in the behavioral sequence as the male approaches the female. It is important to know the role of all pheromone components and to know which part of the precopulatory behavior is the most vulnerable to disruption.

Role of Pheromone Components

The answers, in part, must come from a return to studying the insects in nature. It is like returning to the days of Fabre, equipped with modern equipment (such as movie cameras) and with all the accumulated knowledge of the entomologists, chemists, and neurophysiologists. In our initial studies along these lines we switched from the redbanded leafroller moth to the oriental fruit moth, *Grapholitha molesta*. The latter species has an attractant system (Roelofs and Cardé 1974 and references therein) very similar to that of the redbanded leafroller, in that it utilizes a precise mixture of *cis* and *trans* acetates (*cis*-8-dodecenyl acetate [7% *trans*]) along with a third component (dodecyl alcohol for OFM compared to dodecyl acetate for RBLR).

In brief, field observations (Cardé, Baker, and Roelofs 1975) revealed that males only oriented toward the chemical source from a distance if both the *cis* and *trans* compounds were present in about a 93:7 ratio. No males were seen approaching the traps containing either chemical alone. The utilization of only the geometrical isomers for long-distance behavior makes the "active space" (Wilson 1970) of pheromone transmission much easier to calculate and rationalize. The addition of 12:OH did not seem to increase the number of males approaching the trap, but did effect an increase in the number of males landing on the trap surface. Also, 12:OH increased the frequency of (a) wing fanning while walking, (b) approaches to the chemical sources, and (c) extrusion of hairpencils (fig. 12). This is not only the

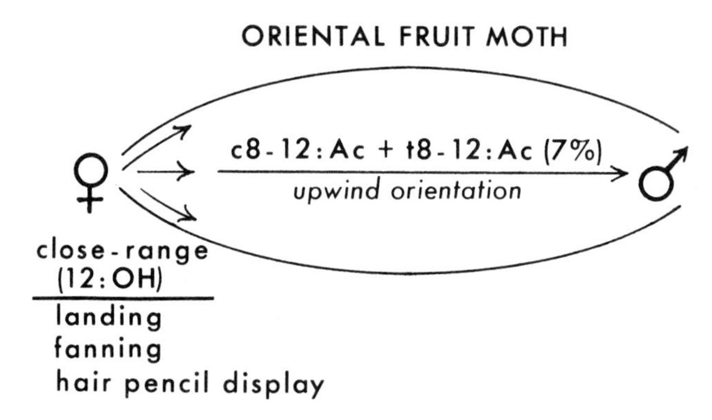

Fig. 12. Proposed role of the attractant components of G. molesta *in producing the precopulatory sequence of male behavioral responses.*

first record of the occurrence of hairpenciling in the Tortricoid superfamily, but also the first case in which an attractant-system component not involved in upwind orientation has been shown to evoke the hairpencil display. In the present example it is not clear which chemicals could best be used to disrupt the mating process.

Only a few of the challenges that entomologists can find in the field of insect communication have been mentioned here. The examples serve to emphasize that there must be continual interaction among a number of related disciplines. It should be evident that there are many important questions that can only be answered by the "enlightened entomologists"—those who study the insects with the genius and keen observation of past masters, equipped with a continual awareness of the pertinent new techniques, instrumentation, and knowledge from their own and related fields of science.

REFERENCES

Birch, M.C. [ed.]. 1974. Pheromones. Elsevier Publ. Co., New York.

Butenandt, A., R. Beckmann, D. Stamm, and E. Hecker. 1959. Über den Sexuallockstoff des Seidenspinners *Bombyx mori*. Reindarstellung und Konstitution. 2. Naturforsch. 146:283-84.

Cardé, R., T. Baker, and W.L. Roelofs. 1975. Behavioural role of individual components of a multichemical attractant system in the Oriental fruit moth. Nature 253:348-49.

Cardé, R., A. Comeau, T. Baker, and W. Roelofs. 1975. Moth mating periodicity: temperature regulates the circadian gate. Experientia 31:46-48.

Carson, R. 1962. Silent spring. Houghton Mifflin Co., Boston.

Chapman, P.J., and S.E. Lienk. 1971. Tortricid fauna of apple in New York. Special Publication of the New York State Agricultural Experiment Station, Geneva.

Evans, D.A., and L.L. Green. 1973. Insect attractants of natural origin. Chem. Soc. Rev. 2:75-97.

Fabre, J.H. 1916. The great peacock, p. 246-78. *In* The life of the caterpillar. Boni & Liveright, New York.

Feng, K.C. 1972. Morphological, histological, and ultrastructural studies of the female sex pheromone gland and its development in *Argyrotaenia velutinana* (Lepidoptera: tortricidae). Ph.D. dissertation, Cornell University.

Hill, A., R. Cardé, H. Kido, and W. Roelofs. Sex pheromones of the orange tortrix moth, *Argyrotaenia citrana*. J. Chem. Ecol., in press.

Jacobson, M. 1972. Insect sex pheromones. Academic Press, New York.

———. 1974. Insect pheromones, p. 229-76. *In* M. Rockstein [ed.], The physiology of insects, vol. 3. Academic Press, New York.

Kaissling, K.E. 1971. Insect olfaction, p. 351-431. *In* L. Beidler [ed.], Handbook of sensory physiology, vol. 4, Chemical senses. Springer-Verlag, New York.

Klun, J., O. Chapman, K. Mattes, P. Wojtkowski, M. Beroza, and P. Sonnet. 1973. Insect sex pheromones: minor amount of opposite geometrical isomer critical to attraction. Science 181:661-63.

Law, J.H., and F.E. Regnier. 1971. Pheromones. Annu. Rev. Biochem. 40:533-48.

MacConnell, J.G., and R.M. Silverstein. 1973. Recent results in insect pheromone chemistry. Angew. Chem. (Int. Ed.) 12: 644-54.

O'Connell, R. 1972. Responses of olfactory receptors to the sex attractant, its synergist and inhibitor in the red-banded leaf roller, *Argyrotaenia velutinana*, p. 180-86. *In* D. Schneider [ed.], Proceedings of the Fourth

International Symposium on Olfaction and Taste, Seewisen, Germany, 1971. Wissenschaftliche Verlagsgesellschaft MBH, Stuttgart.
———. Olfactory receptor responses to sex attractants in the red-banded leaf roller. Unpublished ms.
Payne, T.L. 1974. Pheromone perception, p. 35-61. *In* M.C. Birch [ed.], Pheromones. Elsevier Publ. Co., New York.
Plummer, E., and R. Silverstein. 1974. Methods for determining the enantiomeric purity of insect pheromones. Abstr. 168th American Chemical Society National Meeting, Atlantic City, Div. Pest Chem., no. 60.
Priesner, E. 1973. Artspezifität und Funktion einiger Insektenpheromone. Fortschr. Zool. 22(1):49-135.
Roelofs, W. Manipulating sex pheromones for insect suppression. Environ. Lett., in press.
Roelofs, W., and H. Arn. 1968. Sex attractant of the red-banded leaf roller moth. Nature (London) 219:513.
Roelofs, W., and R. Cardé. 1974. Oriental fruit moth and lesser appleworm attractant mixtures refined. Environ. Entomol. 3:586-88.
Roelofs, W., and A. Comeau. 1968. Sex pheromone perception. Nature 220:600-601.
———. 1971*a*. Sex attractants in Lepidoptera, p. 91-114. *In* A.S. Tahori [ed.], Chemical releasers in insects. Gordon & Breach, New York.
———. 1971*b*. Sex pheromone perception: synergists and inhibitors for the red-banded leaf roller attractant. J. Insect Physiol. 17:435-49.
———. 1971*c*. Sex pheromone perception: electroantennogram responses of the red-banded leaf roller moth. J. Insect Physiol. 17:1969-82.
Roelofs, W., A. Comeau, A. Hill, and G. Milicevic. 1971. Sex attractant of the codling moth: characterization with electroantennogram technique. Science 174:297-99.
Roelofs, W., A. Hill, and R. Cardé. Sex pheromone components of the redbanded leafroller moth, *Argyrotaenia velutinana*. J. Chem. Ecol., in press.
Roelofs, W., J. Kochansky, R. Cardé, H. Arn, and S. Rauscher. 1973. Sex attractant of the grape vine moth, *Lobesia botrana*. Bull. Soc. Entomol. Suisse 46:71-73.
Schneider, D. 1974. The sex-attractant receptor of moths. Sci. Amer. 231(1):28-35.
Shorey, H.H. 1973. Behavioral responses to insect pheromones. Annu. Rev. Entomol. 18:349-80.
———. 1974. Environmental and physiological control of insect sex pheromone behavior, p. 62-80. *In* M.C. Birch

[ed.], Pheromones. Elsevier Publ. Co., New York.
Tette, J.P. 1974. Pheromones in insect population management, p. 399-410. *In* M.C. Birch [ed.], Pheromones. Elsevier Publ. Co., New York.
Trammel, K., W. Roelofs, and E. Glass. 1974. Sex pheromone trapping of males for control of redbanded leafroller in apple orchards. J. Econ. Entomol. 67:159-64.
Von Siebold, C.T. 1837. Fernere Beobachtung über die Spermatozoen der wirbellosen Thiere. IV. Die Spermatozoen in den befruchteten Insekten-Weibchen. Arch. Anat. Physiol., p. 392-439.
Wilson, E.O. 1970. Chemical communication within animal species, p. 133-55. *In* E. Sondheimer and J. Simeone [eds.], Chemical ecology. Academic Press, New York.

PART III

Patterns and Processes in Insect Interactions

INSECT DISPERSAL

John S. Kennedy

This essay is focused upon what has been perennially the most controversial aspect of insect dispersal, the involvement of specialized behavior. Slow, diffusive dispersal occurs during "trivial" movements within the habitat, of course, but how far can we identify special behavioral changes which ensure *migration*, meaning movements of another order, between different habitat localities? It was for long supposed that many major displacements of insect populations were merely accidental extensions of trivial movements or "vagrancy," "a passive distribution of individuals by overpowering forces such as strong winds" (Williams 1958); hence the continued use of the term *dispersal* alongside *migration* even at the individual level of study.

The study of such dispersals at the level of population ecology is a different problem which falls squarely within Southwood's paper in this symposium. But it is important for my narrower theme to recall how Johnson (1960, 1963, 1969) Southwood (1962, 1972; Southwood et al. 1974), and Dingle (1972, 1974) have recently put new order into our ideas about the ecological role of migration. It has been pointed out that migratory insects are polarized toward the "r" end of MacArthur's "r-K" spectrum (MacArthur 1960; MacArthur and Wilson 1967). This spectrum represents two opposite directions of natural selection of life-history strategies. K stands for carrying capacity, and "K-strategists" live near the carrying capacity of their relatively unchanging environments; r stands for maximal rate of increase, and "r-strategists" are opportunistic colonizers of relatively short-lived environments and are able to multiply rapidly, so that they jerk up and down the steep part of the logistic curve of population growth. Migrants are r-strategists that make their living by traveling from one temporary habitat locality to another, often with heavy losses but at a stage when their "reproductive value" is high.

If these authors are right, it can no longer be doubted that traveling between disconnected breeding sites, diapause sites, and so on, is very much an "evolved adaptation," as

C.G. Johnson (1960, 1963, 1969), with emphatic redundancy, expressed it. It is an integral part of the life strategy of those insects that do it, whether or not they travel regularly to and fro between geographically distinct areas and regardless of the manner of their displacement or the stage in the life history at which it occurs. That is my excuse for now dropping the term *dispersal*, which carries no necessarily adaptive connotation, in favor of *migration* which, whatever else it may mean to different people, does carry adaptive, biological overtones. Natural selection being what it is, I doubt that there is a significant amount of dispersal, catastrophes aside, which is not adaptive.

Starting out, then, from the conception of migration as adaptive traveling, it would seem quite extraordinary if there had not also evolved various behavioral specializations subserving migration, again no matter what the particular biological circumstances. And yet the idea of involuntary dispersal "dies hard," as Johnson wrote in his great monograph of 1969; and indeed it still won't lie down. Plant seeds undergo involuntary dispersal, but we now know that the displacements even of the feeblest insects, thrips, aphids, leafhoppers, and small members of many other orders, do depend on a specialized behavior pattern. They, too, can therefore be characterized as migrants, not only ecologically but also behaviorally.

Often their migratory behavior includes a temporarily dominating positive phototaxis (Johnson 1969) which takes them right out of their "boundary layer" (Taylor 1958, 1960, 1974), defined as the layer of retarded wind flow near the ground within which the insect's airspeed exceeds the wind speed. Above this the insect can no longer prevent carriage by the wind and goes with it. At first sight this mode of travel looks so different from the self-steered, self-propelled migration within the boundary layer that is typical of various strong fliers such as butterflies, that serious doubts have been raised again and again that migration can be defined behaviorally (Rainey 1958; Johnson 1960; Taylor 1974).

In an attempt to bring these two contrasting modes of displacement together, it has been argued (Kennedy 1961*b*) that they both effectively straighten out the flier's track over the ground so that it travels, in the sense of traversing new ground instead of frequently changing its direction in a series of "trivial," station-keeping flights. This straightening out is a common feature of migration and is of fundamental ecological importance, but admittedly it is not a strictly behavioral feature.

A strictly behavioral feature that has been emphasized for a long time is persistent locomotor activity of one kind or another. Such activity, although not peculiar to migrants (for it is seen also in nonmigrating, swarming Nematocera, for example), is admittedly characteristic of many migrants, including the wind-riding fliers. But it is not universal, being absent in small, wind-borne, apterous travelers such as first-instar gypsy-moth caterpillars.

A third feature of migratory behavior that has gained increasing attention in recent years (Kennedy 1961*b*; Johnson 1969; Caldwell 1974; Caldwell and Rankin 1974) is an essentially negative one. This is the temporary depression of responses to those stimuli which would arrest or deviate the movement of the migrating insect, but which dominate the same insect's behavior during its trivial movements within one locality. These are the responses of larvae or adults to stimuli signaling food, of adult females to stimuli signaling oviposition sites, and even of males to sexual signals from females. Such stimuli have been loosely dubbed "vegetative" stimuli because they signalize the local breeding habitat and are generally (although not rigidly) associated with resources for somatic or reproductive growth. The responses to these signals are similarly labeled "vegetative," but it should be noted that this label itself refers to the common outcome of different behaviors and not to any one type.

The depression of such responses is of course characteristic of diapause in insects, and it has now come to be seen as a condition for straightened-out traveling as well. The eventual weakening or disappearance of the behavioral inhibition is what brings the migratory traveling to an end. A simple example is the way the flight activity of certain Scolytid beetles cannot be arrested by the odor of a host tree until after they have been flying for many minutes (Graham 1959; Bennett and Borden 1971).

When referring to this negative feature, authors often characterize the behavior as "undistracted" or "non-appetitive." For the purposes of analyzing behavior, however, these terms are arbitrary and subjective, for they do not identify what kind of stimulus the insect is undistracted by, or what it is that the insect has no appetite for. An insect cannot, after all, do any one thing without inhibition of competing responses. When heading for the sky, a young milkweed bug is indeed undistracted by a host plant, but, equally, when heading for a milkweed plant, it is undistracted by the sky. Objectively, an appetite for something means no more than a specific readiness to respond to stimuli associated with it (a low response threshold), and in heading phototactically for the sky the bug is displaying just that. It seems less ambiguous and more informative, therefore, to say that migrants are undistracted by "vegetative" stimuli, specifically, for it is the "vegetative" responses that tend to be inhibited.

Those three interconnected features—persistent locomotion, the straightening out of movements, and the depression of vegetative responses—are, then, the ones that have been brought forward in recent years as diagnostic of migratory behavior in insects; but there have been considerable reservations about them. Let us look briefly at some contrasting examples of insect migration to see how far they do fit the scheme. These examples will not be restricted to migration by flight because the life-strategy ecologists make no such restriction, nor can we do so, without prejudice, at the behavioral level.

First, let us take Schneirla's (1945, 1957) classic description of pedestrian migration in an army ant, *Eciton hamatum*. These ants make simple baglike "bivouac" nests with their own massed bodies above ground, from which they make two different types of excursion. In one type only workers move out, in columns, along repeatedly branching scent trails laid by those momentarily out in front, and they seize, dismember, and carry back to the nest all small life unlucky enough to lie in their way. They all return to the nest in the evening. This is plainly foraging activity, in which the ants are very readily deviated, arrested, and turned back toward the nest again by prey objects used for colony growth.

In the second type of excursion, however, the entire colony is entrained, and the excursion ends with the setting up of a new bivouac nest at a new site. This is not just an extended foraging trip. It is apparently brought on by the maturing of a new brood within the colony which has a powerfully stimulating effect on the locomotor activity of the ordinary mature workers. More of them now go foraging during the day, and towards evening their return movement reverses into a second exodus that carries the whole colony away—queen, brood, and all—to a new bivouac site. This movement is directed along only one of the scent trails that were made during the day, and now the ants do not react to prey and do not branch off.

Thus the shift of the colony to a new foraging area occurs because of intensified and straightened-out locomotor activity and depends on inhibition of the predatory responses to "vegetative" stimuli that at other times arrest and divert locomotion.

Another example of migration on foot, this time by larvae of all stages, is the so-called marching of young desert locusts, nicknamed "hoppers." It is well known that the behavior of these insects depends upon how crowded they, and their mothers, have been. Previously uncrowded hoppers live quietly and tend to repel one another when they meet. They are found in a well-defined type of semidesert habitat, green after rain but with bare patches where the hoppers bask in the sun. At other times they climb and feed intermittently on the plants. Their movements are variable in direction, and they show a strong visual attraction to upstanding objects which prevents wandering away from the clumped vegetation onto open ground.

Hoppers with a history of continuous crowding, however, are more active than solitary ones even when quite alone, and they strongly attract and activate one another, forming bands of up to tens of thousands of individuals which march for hours each day, day after day. When marching, they are not attracted to upstanding objects but rather dodge round them, staying in the open, and they are arrested only briefly by food plants. On open, flat ground the path of these marching hoppers is remarkably straight, mainly because of what has been called "gregarious inertia" but also because of a sun-compass reaction which keeps an individual on the same course during temporary separations from its fellows.

So far, then, although the sensory cues involved are different, our two cases both show intensified locomotor activity, some actively oriented straightening out of the path of locomotion, and some inhibition of normal responses to vegetative stimuli.

Next let us consider a species of aphid, a *Pemphigus* living on the roots of the sea-aster plants that grow on mud banks in tidal salt marshes. My information comes from an unpublished thesis by W.A. Foster (1974), and I am grateful for his permission to mention it because it bridges the gap, behaviorally, between the terrestrial, pedestrian migrants and the airborne, wind-propelled ones such as winged aphids. The migrating aster aphid is also wind propelled, but not air-borne. The alate morph is produced in very small numbers, and the principal migratory stage is the first-instar larva, less than 1 mm long.

The other larval instars, and the apterous adults, are consistently photonegative and stay down in cavities round the aster roots where they feed. But the first instar is more active and becomes photopositive, leaving the roots and climbing up to wander on the exposed soil surface. The rising tide lifts it and drifts it, standing erect on the water surface, and exposes it to the wind, which sends it scudding away across open water to be deposited by the falling tide on another mud bank. Thirty minutes of this waterborne experience is enough to reverse the reaction of these larvae to light and send them down a new crevice where aster roots may be found and colonized; whereas, on the soil, many hours are required to produce the same reversal to photonegative behavior. This makes a striking parallel with the ambrosia beetles (Graham 1961; Francia and Graham 1967) and other insects in which airborne experience causes a reversal from positive to negative phototaxis.

But we notice here a radical behavioral difference from the terrestrial, pedestrian migrants, as well as from my next case of a flying one that keeps low within its boundary layer. The straightening out of the horizontal track of the aster-aphid larvae, like that of small moth larvae that let themselves go on silk threads in the wind, is not an active orientation response. It is merely an indirect consequence of the specialized migratory behavior pattern, which consists in leaving the host plant and by positive phototaxis getting into a position that ensures wind-propelled travel, while the reversing phototaxis provides some indirect control over the distance traveled.

Migration by flight is surely the most usual mode of migration in insects, and as a first case of it let us take the only one where we have a full, start-to-finish description of migration entirely within the boundary layer, that of the Pierid butterfly, *Ascia monuste*, on the Florida coast as described by Nielsen (1961). Within the larval host-plant habitat, the behavior of the adults consists of flying hither and thither, with stops to feed, bask, clean, and, with sexual maturity, to mate and lay eggs. Each morning they make local flights from the breeding site to nearby flower patches where they feed on the nectar. Under certain conditions, when there are large numbers of butterflies doing this, their "flits" between feeding sites become more and more concerted and sustained. Turnings aside to stop and feed become briefer and further apart until there emerges a solid stream of butterflies all going the same way, rarely turning aside or stopping, and covering tens of miles. Eventually they settle back into normal localized living in a new habitat, thanks to the reappearance of the responses to plants, which were largely inhibited during the migratory flight.

Here the traveling from one habitat locality to another is self-propelled and self-steered, like that of the pedestrian migrants. The ground track seems to owe nothing to the wind; the fliers react to an increase of wind speed by flying lower in relative shelter, actively keeping within their boundary layer. There may be separate streams going in different directions in the same area at the same time (as with locust hoppers), and the straightening out is not directed to any particular place but results simply from stabilizing, visual orientation responses to the shoreline and other gross features such as roads and sand dunes and, very probably, the sun (Baker 1968*a* and *b*).

Next let us go to the other extreme and look at an example of migration by flight above the boundary layer by one of the smaller insects which have airspeeds rated in centimeters rather than meters per second. In the black bean aphid, *Aphis fabae*, the situation is, in principle, remarkably similar to that in the minute aster-aphid larvae. The oustanding difference is that there is now persistent locomotor activity. Without their continued wing-beating, these insects would not remain airborne for long, since the sinking speed of a bean aphid with its wings spread but not beating is about 80 cm/sec. The salient behavioral features of the migration of this aphid (Johnson 1969; Dixon 1971; Kring 1972; Kennedy and Fosbrooke 1972) may be summarized as follows.

The young adult completes the teneral stage with an internal inhibition of the responses to normally arresting stimuli from the host plant, and is instead stimulated by light to take off from the plant. This inhibition is progressively removed during flight but can be reversibly reimposed by landings on nonhost plants. There are accompanying changes in the phototactic responses, more sophisticated than the simple positive-negative reversals in the aster-aphid larva. The flier is first drawn up out of its boundary layer by the sky, thus becoming subject to the atmospheric circulation, but later it is drawn down again by the sight of the green earth. In this way it reenters its boundary layer and there makes test alightments on plants until it perishes or finds a host.

The migratory behavior of many medium-sized insects, too, such as moths and mosquitoes, has long been known to involve rising out of the boundary layer to travel with the wind (Johnson 1969), and it was a major blow to the earlier idea that insect migrants typically steer themselves during their travels when Rainey (1951, 1963) demonstrated beyond doubt that day-flying insects as large and powerful as the locusts, with airspeeds of over 4 m/sec, also traveled routinely downwind. In view of this, and in the absence of any unequivocal evidence to the contrary, it seemed that large butterflies like the monarch might also achieve their long-range displacements on the wind, and need not steer themselves all the way inside their boundary layer as had long been believed. Entomologists even suggested that birds, too, might migrate in this way (Rainey 1960).

Now, however, the picture has changed once again. Insects flying above their boundary layer cannot, by definition, steer themselves on an upwind or cross-wind ground track because the wind is too strong at that level. But it has now become clear that that does not justify the usual assumption that the straightening out of their ground tracks is done entirely by the wind. Their own active behavior can make a significant contribution to their horizontal travel, not only through staying aloft but also through their orientation and their thrust.

Fifteen years ago, when Klassen and Hocking (1964) were watching young migratory adults of the mosquito *Aedes cataphylla* soaring up into the free-moving air in the evening and then flying away downwind as so many migrants do, they noticed that the mosquitoes were actually turning and heading downwind from the moment they began to be carried that way. By doing so, they increased their groundspeed substantially, adding their own airspeed to the weak night wind (cf. Roffey 1963).

More recently, Waloff (1972) has demonstrated the same thing in day-flying desert locusts, by means of photographs taken vertically upwards into dense swarms undergoing a net downwind displacement. Although the lower locusts showed a confusion of orientations and ground tracks, as well as much landing and taking off again, she found that the upper locusts were consistently headed downwind. Adding their own considerable airspeed of 5 m/sec to the wind speed, they raced ahead of the lower fliers until they passed out in front, whereupon they came down and turned back into the main mass. Thus those fliers that were for the time being higher in the air appeared to be making the running and effectively setting the direction of displacement of the swarm as a whole. What sensory cues the high-flying locusts are using in orienting downwind we do not know, but Waloff gives grounds for believing that the orientation to wind could be optomotor; that is, a response to the apparent movement of the ground pattern that is caused by the wind carrying the insect.

This subject has now been thrown wide open by the radar observations of Schaefer (1969, 1972, and in press), Roffey (1969, 1972), and Riley (1974, 1975). Nonswarming locusts, grasshoppers, crickets, noctuids, tortricids, sphingids, and other insects have been discovered flying at night in greater numbers and diversity than by day, in West and East Africa, Australia, and Canada. We still have only summary accounts of the great mass of results obtained by Schaefer especially, but some of the reported behavioral findings cannot pass unmentioned.

In the daytime the role of behavior in the vertical distribution of insects that migrate above their boundary layer is confused, for us, by the convective lift that is usually available and gives their vertical-density profile an impressive resemblance to that of inert particles (Johnson 1969). There is no such ambiguity at night, because the insects seen on the radar screen are usually rising through a temperature inversion with no help from convection. They rise to ceilings at anything from a few hundred meters to 1-2 km, depending on temperature, with the density profile varying between species and occasions, thus confirming the active role of behavior.

The ground tracks of these night migrants were typically downwind, as ground observers have often reported before (Johnson 1969). But there was a hint that the migrants might choose, as it were, their direction of displacement by flying at the appropriate height. For example, moths and grasshoppers were seen traveling in opposite directions in the lower and upper winds, respectively, near the intertropical front in the Sudan. But, most astonishingly, the observers agree that the fliers, although sometimes oriented at random, often show a common orientation that is close to downwind, among the overwhelming majority of individuals on the radar screen, up to hundreds of meters. In his very extensive experience, Schaefer says he found uniformity of orientation, with a good component in the downwind direction, on most occasions. The result is that the fliers' groundspeeds exceed the speeds of the winds carrying them by several meters per second.

The riddle is what environmental cues they are using for their orientation. Schaefer mentions occasions when spruce-budworm moths maintained their downwind orientation at night under complete overcast or between two cloud layers. If this is confirmed, we shall have to follow the ornithologists, who have many well-documented records of consistent orientation under such handicaps, and have been looking for quite new kinds of sensory guidance: geomagnetic, acoustic, or involving the fine structure of the air movement itself (Gauthreaux 1972; Griffin 1973; Griffin and Hopkins 1974).

What about the optomotor anemotaxis, assuming for the moment that there is enough light on most nights? Optomotor anemotaxis, as already mentioned, depends on the flier responding to the apparent movement of the ground pattern due to wind drift, and one might think this movement would be much too slow to be detected by high fliers. For the locust Burtt (in Waloff 1972) gives a threshold retinal velocity of about 4°/sec. Given that sensitivity, a locust could, theoretically, detect lateral drift due to winds of 5-10 m/sec at heights of 75-150 m above ground level. Waloff (1972) suggests that day-flying swarm members seen rising to still greater heights could be maintaining the orientation that they had acquired lower down, by continued gregarious alignment on their fellows or a sun-compass reaction.

The rigorous experiments of Thorson (1964) and Kien (in press) suggest much lower retinal-velocity thresholds in locusts, which would theoretically permit lateral drift detection up to thousands of meters above ground. But such height estimates seem to require the impossible—a rock-steady insect eye in a laminar airflow; moreover, the night fliers are usually too far apart for likely gregarious reactions. So we must, as C.B. Williams believed, start looking into some sensory guidance mechanisms not yet considered for insects.

However that may be, it is plain that we can no longer assume that actively oriented traveling is confined to the minority of insects that migrate within their boundary layer near the ground. The behavior of many migrants plays a major part in straightening out their ground tracks and in regulating the distances traveled.

As to their control of direction, it has long been known that a number of north-south continental displacements of insect populations, with particular seasonal winds, are ecologically advantageous (Johnson 1969; Bowden 1973; Hughes and Nicholas 1974). For the spruce-budworm moths there seems to be no particular direction that would be advantageous, and their observed downwind orientation at night probably serves only to lengthen their trajectories. However, it has seemed incredible that simply heading downwind could always be advantageous for all migratory species, and Schaefer (in press) now claims to have radar evidence of something approaching true navigation by night-flying grasshoppers in the Sudan. At a season when their next habitat areas lie to the SSW, they fly downwind in great numbers in the strong trade winds blowing that way, but in light winds blowing in other directions they deviate from downwind in such a way as to travel roughly on the same compass bearing toward the SSW. Those that fly when there is a strong monsoon wind blowing toward the NE again fly downwind, which is the "wrong" direction, but then very few are seen flying at all.

What can we conclude, from this anecdotal survey, about the behavior of insect migrants in general? Firstly, that there has indeed been an evolutionary differentiation of behavior securing adaptive displacements of all kinds. Secondly, that the departure and the groundspeed, the direction and the distance are all determined by behavior to much greater extent than was thought until very recently. Thirdly, that this behavioral specialization does show certain features common to insect migrants generally, but that these features are much broader in character than was once thought. Insect migration can be defined behaviorally.

Two of the three behavioral features that were presented as common—namely, persistent locomotor activity and oriented straightening out of the ground track—are both very general, but obviously not universal. It seems very unlikely that small fliers like aphids orient themselves downwind, and the wingless wind-riders of course cannot.

Nevertheless, further work may show that there is a common feature in the orientation behavior of migrants, whether or not this is directly responsible for the straightening out of the ground track. The common element here would lie in the class of orienting cue from the environment to which migrants are for the time being most responsive. Pedestrian and flying migrants have been described as relatively unresponsive to the transient type of stimulus that comes from a single, localized environmental feature like a single plant, and especially responsive to continuing stimuli such as come from gross, pervasive features like sun, wind, and relief, with which the migrant tends to stabilize its sensory relations (Kennedy 1951).

But that feature of migrant behavior would be only a particular aspect of what was put forward as the third common behavioral feature, some temporary inhibition of responses to sensory inputs that signal vegetative resources. In one form or another, this seems to be the surest way of distinguishing migratory from other kinds of behavior—although of course we are always dealing with a complex continuum, not an absolute category.

There is also a transient depression of vegetative physiological functions accompanying the depression of behavioral responsiveness to the vegetative signal stimuli. In females migratory flying is usually prereproductive (or interreproductive), in a period of arrested or retarded oocyte development (Johnson 1969; Dingle 1972, 1974; Waloff 1973). Compared with nonmigratory females of the same or related species, migratory ones take longer to start laying eggs and lay fewer. This once provided the basis for a general hypothesis that migratory flying is, like adult diapause in many insects, part of a juvenile hormone-deficiency syndrome (Kennedy 1956; Johnson 1963; cf. Southwood 1961). This seemed an attractive unifying hypothesis, but it has come up against accumulating evidence (see Truman and Riddiford 1974) that juvenile hormone can promote locomotor activity, including flight activity in migratory locusts.

The fullest evidence comes from the work of Caldwell and Rankin (1972; Rankin 1974) on the milkweed bug, *Oncopeltus fasciatus*, which migrates during the extended period of sexual immaturity that is induced by short days or incipient starvation. Both natural and synthetic juvenile hormone greatly increase the proportion of bugs making long flights, males as well as females. This hormone also has its usual effect of promoting egg ripening in these females and, after some days, it terminates their sustained flights by some unidentified feedback from the ovaries.

How can the same hormone have such conflicting effects? Caldwell and Rankin (1972) first suggested, and now Rankin (*in litt.*) has proved, with the aid of the very sensitive *Manduca*-assay method for juvenile hormone, that the natural hormone is present at moderate levels in the blood during the days of sustained flights and then rises further, and it is not until it reaches the higher level that the ovaries respond, the flight activity wanes, and oviposition supervenes. So it is still true that long flights by this bug depend upon a *relative* deficiency of juvenile hormone. But a moderate amount of it in the blood not only permits such flights, it positively promotes them.

Yet juvenile hormone does not seem to have that effect in the few other bugs and beetles where it has been tested, and in some it even induces breakdown of the flight muscles (see Rankin 1974). Altogether, it looks now as if we shall not find any common physiological mechanism in migration, even in migration by flight only. Probably we cannot expect to see much order in the different ways that the endocrine and other physiological systems are used in migration until we can place those different ways in the wider concept of life strategies, and this is something Rankin (1974) has begun to do.

Meanwhile all this poses a new question about life strategies. If migration does involve some deferment of vegetative physiological functions, this seems to contradict the ecologists' thesis with which we began; namely, that migrants are *r*-strategists specialized for rapid reproductive turnover and population growth. Migrants put less of their matter and energy into reproduction and more into maintenance (by migration) than do comparable nonmigrants, which is just the opposite of what an *r*-strategist "should" do (Pianka 1970). In other words, going from a less to a more migratory state in the individual or the population is not a move towards the *r* end of the *r*-*K* spectrum of life strategies, but a move towards the *K* end.

The contradiction is more apparent than real, however, because what the ecologists claim is that the migratory *r*-strategist has been selected for "rapid population growth at very low densities when a few individuals have invaded an ecological vacuum" (Southwood et al. 1974). The deferment of reproduction until that vacuum situation has been reached is a physiological condition first for reaching it and then for profiting from its "vegetative" advantages. The individual entering upon its transient migratory phase, and likewise the population going over to the production of migratory instead of nonmigratory forms, are indeed taking a temporary step back towards the *K* end of the spectrum, and it is by just this tactic that they maintain their undoubted position near the *r* end.

What is today called the *r*-*K* axis seems to be functionally the same as what was some years ago clumsily called the "vegetative-sensorimotor" axis; and that, so the argument ran, is the same again as the juvenile-adult axis, because "adults find growth resources, while juveniles exploit them." Thus an individual is at its most adult not when it is actually reproducing, but rather during that period of physiological reversal wherein the products of one phase of growth are not added to but employed behaviorally in ensuring the success of the next (Kennedy 1956, 1961*a* and *b*; Kennedy and Stroyan 1959).

Plants have dispersal mechanisms, but plants do not have locomotory behavior. The evolution of animals as a kingdom displays the vast opportunities for adaptive radiation that have flowed from periodically foregoing a "vegetable" existence and launching upon some energy-expending activity that brings the animal into contact with fresh resources elsewhere. Insect migration is one rather extreme, but very successful, example of that peculiarly zoological sort of evolutionary compromise.

ACKNOWLEDGMENTS

I am grateful to Drs. W.A. Foster, M.A. Rankin, and J.R. Riley and Prof. G.W. Schaefer for permission to quote their unpublished results, and to Dr. J.N. Brady for criticism.

REFERENCES

Baker, R.R. 1968*a*. A possible method of evolution of the migratory habit in butterflies. Phil. Trans. Roy. Soc. London (B) 253:309-41.

———. 1968*b*. Sun orientation during migration in some British butterflies. Proc. Roy. Entomol. Soc. London (A) 43:89-95.

Bennett, R.B., and J.H. Borden. 1971. Flight arrestment of tethered *Dendroctonus pseudotsugae* and *Trypodendron lineatum* (Coleoptera: Scolytidae) in response to olfactory stimuli. Ann. Entomol. Soc. Amer. 64:1273-86.

Bowden, J. 1973. Migration of pests in the tropics. Meded. Fak. Landbouwwetens. Rijksuniv. Gent. 38:785-96.

Caldwell, R.L. 1974. A comparison of the migratory strategies of two milkweed bugs, *Oncopeltus fasciatus* and *Lygaeus kalmii*, p. 304-16. *In* L.B. Browne [ed.], Experimental analysis of insect behaviour. Springer-Verlag, Berlin.

Caldwell, R.L., and M.A. Rankin. 1972. Effects of a juvenile hormone mimic on flight in the milkweed bug, *Oncopeltus fasciatus*. Gen. Comp. Endocrinol. 19:601-5.

———. 1974. Separation of migratory from feeding and reproductive behavior in *Oncopeltus fasciatus*. J. Comp. Physiol. 88:383-94.

Dingle, H. 1972. Migration strategies of insects. Science 175:1327-35.

———. 1974. The experimental analysis of migration and life-history strategies in insects, p. 329-42. *In* L.B. Browne [ed.], Experimental analysis of insect behaviour. Springer-Verlag, Berlin.

Dixon, A.F.G. 1971. Migration in aphids. Sci. Progr. Oxford 59:41-53.

Francia, F.C., and K. Graham. 1967. Aspects of orientation behavior in the ambrosia beetle *Trypodendron lineatum* (Oliver). Can. J. Zool. 45:985-1002.

Foster, W.A. 1974. The biology of a salt marsh aphid (*Pemphigus* sp.) Ph.D. thesis, University of Cambridge.

Gauthreaux, S.A., Jr. 1972. Flight directions of passerine migrants in daylight and darkness: a radar and visual study, p. 129-37. *In* S.R. Galler, K. Schmidt-Koenig, G.J. Jacobs, and R.E. Belleville [eds.], Animal orientation and navigation. U.S. Govt. Print. Office, Washington.

Graham, K. 1959. Release by flight exercise of a chemotropic response from photopositive domination in a scolytid beetle. Nature, London 184:283-84.

———. 1961. Air swallowing: a mechanism in photic reversal of the beetle *Trypodendron*. Nature, London 191:519-20.

Griffin, D.R. 1973. Oriented bird migration in or between opaque cloud layers. Proc. Amer. Phil. Soc. 117:117-41.

Griffin, D.R., and C.D. Hopkins. 1974. Sounds audible to migrating birds. Anim. Behav. 22:672-78.

Hughes, R.D., and W.L. Nicholas. 1974. The spring migration of the bushfly (*Musca vetustissima* Walk.): evidence of displacement provided by natural population markers including parasitism. J. Anim. Ecol. 43:411-28.

Johnson, C.G. 1960. A basis for a general system of insect migration and dispersal by flight. Nature, London 186:348-50.

———. 1963. Physiological factors in insect migration by flight. Nature, London 198:423-27.

———. 1969. Migration and dispersal of insects by flight. Methuen & Co., London.

Kennedy, J.S. 1951. The migration of the Desert Locust (*Schistocerca gregaria* Forsk.). I. The behaviour of swarms. II. A theory of long-range migrations. Phil. Trans. Roy. Soc. London (B) 235:163-290.

———. 1956. Phase transformation in locust biology. Biol. Rev. 31:349-70.

———. 1961*a*. Continuous polymorphism in locusts. Symp. Roy. Entomol. Soc. London 1:80-90.

———. 1961*b*. A turning point in the study of insect migration. Nature, London 189:785-91.

Kennedy, J.S., and I.H.M. Fosbrooke. 1972. The plant in the life of an aphid. Symp. Roy. Entomol. Soc. London 6:129-44.

Kennedy, J.S., and H.L.G. Stroyan. 1959. Biology of aphids.

Annu. Rev. Entomol. 4:139-60.
Kien, J. Sensory integration in the locust optomotor system. I. Behavioural analysis. II. Direction-selective neurons in the circumoesophageal-connectives and the optic lobe. Vision Res. 14, in press.
Klassen, W., and B. Hocking. 1964. The influence of a deep river valley system on the dispersal of *Aedes* mosquitoes Bull. Entomol. Res. 55:289-304.
Kring, J.B. 1972. Flight behavior of aphids. Annu. Rev. Entomol. 17:461-92.
MacArthur, R. 1960. On the relative abundance of species. Amer. Natur. 94:25-34.
MacArthur, R., and E.O. Wilson. 1967. The theory of island biogeography. Princeton University Press, Princeton.
Nielsen, E.T. 1961. On the habits of the migratory butterfly *Ascia monuste* L. Biol. Medd. 23:1-81.
Pianka, E.R. 1970. On *r*- and *K*-selection. Amer. Natur. 104: 592-97.
Rainey, R.C. 1951. Weather and the movement of locust swarms: a new hypothesis. Nature, London 168:1057-60.
———. 1958. Biometeorology and the displacements of airborne insects. First Int. Bioclimatol. Congr., Vienna, 1957, 3(B).
———. 1960. Applications of theoretical models to the study of flight-behaviour in locusts and birds. Symp. Soc. Exp. Biol. 14:122-39.
———. 1963. Meteorology and the migration of Desert Locusts. Locusts. Tech. Notes World Meteorol. Org. 54:115.
Rankin, M.A. 1974. The hormonal control of flight in the milkweed bug, *Oncopeltus fasciatus*, p. 317-28. *In* L.B. Browne [ed.], Experimental analysis of insect behaviour. Springer-Verlag, Berlin.
Riley, J.R. 1974. Radar observations of individual desert locusts (*Schistocerca gregaria* (Forsk.))(Orthoptera, Locutidae). Bull. Entomol. Res. 64:19-32.
———. 1975. Collective orientation in night-flying insects. Nature, London 253:113-14.
Roffey, J. 1963. Observations on night flight in the Desert Locust (*Schistocerca gregaria* Forskal). Anti-Locust Bull. 39.
———. 1969. Report on radar studies on the Desert Locust *Schistocerca gregaria* (Forskal), in Niger Republic September-October 1968. Anti-Locust Occas. Rep. 17.
———. 1972. Radar studies of insects. PANS 18:303-9.

Schaefer, G.W. 1969. Radar studies of locust, moth and butterfly migration in the Sahara. Proc. Roy. Entomol. Soc. London (C) 34:33, 39-40.

———. 1972. Radar detection of individual locusts and swarms, p. 379-80. *In* C.F. Hemming and T.H.C. Taylor [eds.], Proceedings of the International Study Conference on the Current and Future Problems of Acridology, London, 1970. Centre for Overseas Pest Research, London.

———. Radar observations of insect flight. Symp. Roy. Entomol. Soc. London, in press.

Schneirla, T.C. 1945. The army-ant behavior pattern: Nomad-statary relations in the swarmers and the problem of migration. Biol. Bull. 88:166-93.

———. 1957. Theoretical consideration of cyclic processes in Doryline ants. Proc. Amer. Phil. Soc. 101:106-33.

Southwood, T.R.E. 1961. A hormonal theory of the mechanism of wing polymorphism in Heteroptera. Proc. Roy. Entomol. Soc. London 36:4-6.

———. 1962. Migration of terrestrial arthropods in relation to habitat. Biol. Rev. 37:171-214.

———. 1972. The role and measurement of migration in the population system of an insect pest. Trop. Sci. London 13:275-78.

Southwood, T.R.E., R.M. May, M.P. Hassell, and G.R. Conway. 1974. Ecological strategies and population parameters. Amer. Natur. 108: 791-804.

Taylor, L.R. 1958. Aphid dispersal and diurnal periodicity. Proc. Linn. Soc. London 169:67-73.

———. 1960. The distribution of insects at low levels in the air. J. Anim. Ecol. 29:45-63.

———. 1974. Insect migration, flight periodicity and the boundary layer. J. Anim. Ecol. 43:225-38.

Thorson, J. 1964. Dynamics of motion perception in the Desert Locust. Science 145:69-71.

Truman, J.W., and L.M. Riddiford. 1974. Hormonal mechanisms underlying insect behaviour. Advan. Insect Physiol. 10: 297-352.

Waloff, N. 1973. Dispersal by flight of leafhoppers (Auchenorrhyncha: Homoptera). J. Appl. Ecol. 10:705-30.

Waloff, Z. 1972. Orientation of flying locusts, *Schistocerca gregaria* (Forsk.), in migrating swarms. Bull. Entomol. Res. 62:1-72.

Williams, C.B. 1958. Insect migration. Collins, London.

UTILIZATION OF INSECT-PLANT INTERACTIONS IN PEST CONTROL

Mano D. Pathak

Insect-plant interactions involve cause and effect relationships between phytophagous insects and their host plants. These include various mechanisms of host selection by the insect; insect utilization of the host plant for food, oviposition, and shelter; the effects of insect infestation of a plant; and the effects of the infested plant on insect survival, growth, and population build-up. Insect pests multiply rapidly on suitable hosts, while their populations are constrained by less suitable hosts. Such differences in suitability occur both among and within species of host plants and can serve as an important method of pest control.

The various insect-plant interactions that help to protect crop plants from insect-pest damage are collectively known as *varietal resistance*. Among the definitions of *varietal resistance*, the following definition by Painter (1951) has been accepted as standard by the majority of workers and appropriately deals with the practical aspects of pest control in crop production:

> Relative amount of heritable qualities possessed by the plant which influence the ultimate degree of damage done by the insect. In practical agriculture it represents the ability of a certain variety to produce a larger crop of good quality than do ordinary varieties at the same level of insect populations.

Painter (1951) classified varietal resistance into three broad categories:

1. nonpreference—when a plant possesses characteristics that make it unattractive to insect pests for oviposition, feeding, or shelter;
2. antibiosis—when the host plants adversely affect the bionomics of the insects feeding on them;
3. tolerance—when the host plant undergoes only slight injury in spite of supporting an insect population

adequate to damage severely susceptible hosts.
A plant variety which fits into one or more of these three categories can be classified as resistant.

The word *relative* is important in the definition of *varietal resistance* since it is rare to find host-plant varieties that are immune to insect attack; even varieties considered highly resistant exhibit some damage under heavy insect infestations. Many resistant crop varieties previously studied had only a moderate level of resistance and consequently were damaged by heavy infestations of insects. This created controversy among some earlier workers; in fact, the limited examples of known varietal resistance were often referred to as exceptions, implying that work on varietal resistance as a method of insect control deserved low priority. However, several recent examples of high levels of monogenic resistance and the realization that other methods of insect control have their limitations have prompted investigation in this area. Also, at one time some workers held the notion that host-plant resistance was associated with undesirable agronomic characters. However, it has been clearly demonstrated that desirable characters of crop plants and insect resistance are not mutually exclusive.

The extensive and growing literature on host-plant resistance precludes a full review of the literature in this paper. Instead only a few prominent examples, particularly in rice research, have been selected to illustrate the various points discussed.

Varietal resistance and antibiosis cannot be considered synonymous, because nonpreference is also an important factor. In those situations where brief infestations could cause severe plant damage, such as severing of the growing parts of the plant or transmission of virus diseases, nonpreference may actually be of greater importance than antibiosis. In field plantings nonpreferred crop varieties frequently escape infestations or develop low ones, and insects, when caged on nonpreferred hosts, lay fewer eggs and develop smaller populations. Thus both antibiosis and nonpreference influence insect populations. On the other hand, tolerant varieties do not inhibit insect multiplications; moreover, because they can support larger infestations while sustaining little plant damage, they may actually be more conducive to insect population build-up than susceptible crop varieties. Tolerance is generally attributable to vigorous plant growth. This paper does not include a detailed discussion of tolerance.

The literature on insect resistance in crop plants and on

the causes of resistance has been reviewed by Painter (1951, 1958, 1966); Beck (1965); Luginbill (1969); Maxwell, Jenkins, and Parrott (1972); Thorsteinson (1960); and Pathak (1970, 1972).

CAUSES OF RESISTANCE

Plant resistance to insects is complex and is seldom caused by a single mechanism or a single chemical. While the classification of varietal resistance into nonpreference, antibiosis, and tolerance is simple, these categories deal with effects, not causes. Little information is available as to what are the actual causes of these effects. Often similar results come from diverse causes.

Nonpreference

Detailed knowledge of insect behavior is crucial to an understanding of the nonpreference response of the insect to a particular plant. The insect is attracted to or repelled from a plant by a variety of characters such as its shape, size, color, surface texture, or chemical constituents. Thorsteinson (1960) described the host selection by phytophagous insects as a series of "take it or leave it" situations. He categorized these into food-plant finding factors, permissive factors, tactile stimuli, visual stimuli, chemotactic stimuli, olfactory stimuli, and factors that initiate and sustain feeding-chemotactic influences of saccharides, organic nitrogen compounds, minerals, organic acids, vitamins, plant pigments, glycosides, alkaloids, water, and others. Broadly, the nonpreference response can be separated into nonpreference for oviposition and nonpreference for feeding.

Nonpreference for oviposition. The striped borer moth, *Chilo suppressalis* (Walker), exhibits a strong preference for oviposition on certain rice varieties, with susceptible rice varieties generally receiving ten to fifteen times more egg masses than the resistant varieties (Patanakamjorn and Pathak 1967; Pathak et al. 1971). Even when the borer incidence was low, susceptible varieties received large numbers of egg masses and were heavily damaged, while most resistant rice varieties were almost insect free.

Plant height, width and length of the flag leaves, stem diameter, and smooth leaf-blade surface were all positively correlated with borer susceptibility (table 1). Although, in

TABLE 1

Correlations between rice-plant characters and percentages of tillers infested with striped borer (Pathak et al. 1971)

Plant Character	Correlation Coefficient
Elongated internodes, number	0.632**
Third elongated internode, length	0.715**
Flag leaf, length	0.798**
Flag leaf, width	0.836**
Culm, height	0.796**
Culm, external diameter	
at half its length	0.672**
at one-fourth its length from the base	0.785**
Culm, internal diameter	
at half its length	0.671**
at one-fourth its length from the base	0.790**
Tillers per plant, number	-0.756**
Stem area occupied by vascular bundle sheaths, percentage	-0.756**

general, plant varieties with hairy leaf surfaces received fewer eggs, removal of hairs did not alter the nonpreference of the ovipositing borer moths. Also, many crop varieties which are preferred for oviposition exhibited no distinct morphological difference from several nonpreferred varieties, suggesting that differences in plant morphological characters alone may not be responsible for the nonpreference of the moths for oviposition.

Volatile chemicals emanating from corn foliage affect the orientation of moths of the European corn borer, *Ostrinia nubilalis* (Hübner)(Moore 1928; and Painter 1951). Differences in the oviposition preferences of the cabbage maggot, *Hylemya brassicae* (Bouche), on different varieties have been recorded. Inasmuch as the *H. brassicae* fly oviposits in the soil near the base of the cabbage plant and not on the plant itself, these differences may be due to differences in the attractiveness of the plant to the gravid flies (Radcliffe and Chapman 1960; Doane and Chapman 1962). The wavelength and intensity of light reflected by the plant foliage influence the orientation of gravid corn earworm moths, *Heliothis zea* (Boddie)(Callahan 1957).

These and a few other factors play important roles in bringing the insect on or near the plant host, but tactile proprioceptive, chemotactic, and visual factors are responsible for site selection and subsequent oviposition (Beck 1965).

Nonpreference for Feeding. An insect feeds on a plant in a sequence of stereotyped behavioral steps that closely parallels that described for oviposition. The three indispensable steps in feeding behavior are (a) host-plant recognition, (b) initiation of feeding (biting or piercing), and (c) maintenance of feeding (Beck 1965). A plant that does not induce these steps is nonpreferred.

The brown planthopper, *Nilaparvata lugens* (Stal.), punctures the tissues of the resistant rice variety Mudgo. Its proboscis reaches the proper feeding tissues, but the insect does little feeding. Mudgo contains a much smaller proportion of the amino acid asparagine than do susceptible rice varieties. In separate tests female planthoppers were strongly attracted to this amino acid (Sogawa and Pathak 1970); thus the low level of asparagine in Mudgo may cause the insect's limited feeding on this host. Chemical constituents in various parts of cotton plants act as attractants, repellents, ovipositional arrestants, and feeding stimulants to the boll weevil, *Anthonomus grandis* (Boheman)(Keller, Maxwell, and Jenkins 1962; Keller et al. 1963; Maxwell, Jenkins, and Keller 1963). For the Colorado potato beetle, *Leptinotarsa decemlineata* (Say), the alkaloids tomatin and demissin extracted from solanaceous plants serve as repellent and feeding deterrent, respectively (Wilde 1958).

Antibiosis

Biophysical Basis. The morphological and anatomical characteristics of a plant unquestionably influence its utility as an insect host. However, in some cases such characteristics may not be the only causes of resistance; there may be other complementary factors. For example, the long and tight husks in certain corn varieties have been found not to be the main cause of resistance to the corn earworm, but, because they confine the larvae longer in the husk, they increase the probability of cannibalism among the larvae (Painter 1951) and also force them to feed on silk channels which, in certain genetic lines of corn, contain a biochemical lethal factor (Guthrie and Walter 1961).

The solid-stemmed nature of Rescue and other wheats is known to be the principal cause of resistance to the wheat stem sawfly, *Cephus cinctus* (Norton). In these resistant wheat varieties, the sawfly eggs tend to be mechanically damaged and desiccated, and the hatching larvae are limited in their movements. Wallace, McNeal, and Berg (1973) have investigated the minimum stem solidness required to attain field control of the sawfly. Another outstanding example studied in recent years is the high level of resistance to the cereal leaf beetle, *Oulema melanopus* (Linn), in pubescent wheat varieties (Schillinger and Gallun 1968; Schillinger 1969; and Ringlund and Everson 1968). The beetle lays fewer eggs on pubescent plants, the eggs laid are susceptible to desiccation, and even the mortality of the few larvae that hatch is higher than that of the larvae which hatch on glabrous varieties.

Resistance to the squash vine borer, *Melittia cucurbitae* (Harris), in cucurbit varieties has been attributed to hard woody stems of the squash plant, the vascular bundles of which are closely packed and tough (Howe 1949). The melon leaf mining larvae, *Liriomyza pictella* (Thomson), mining in snap dragon are crushed by rapidly proliferating wound tissues; and in eggplant the tissues surrounding the mines die quickly, thereby confining the larvae in dead tissues (Oatman 1959).

Resistance to the striped rice borer was observed in rice varieties whose leaf sheaths were tightly wrapped around the stem, in varieties whose stems had closely packed vascular bundle sheaths and a large number of sclerenchymatous layers, and in rices whose stems and leaf sheaths contained high amounts of silica; these structures interfered with the boring activity of the larvae (Patanakamjorn and Pathak 1967; Djamin and Pathak 1968; and Sasamoto 1961). The larvae feeding on rice varieties containing high amounts of silica exhibited the typical antibiosis effects and worn-out mandibles (Djamin and Pathak 1968).

Biochemical Factors. Generally, antibiosis between crop plants and insect pests appears to be due to differences in chemical constituents of the plant. These differences may be qualitative or quantitative and may exist only within certain parts of the plant or during certain stages of plant growth. The determination of these factors is intricate. Differences in chemical constituents of resistant and susceptible crop varieties should be bioassayed before they are identified as the cause of resistance; this is often difficult due to lack of information on the nutrition and biochemical feeding behavior of many phytophagous insects. The use of isogenic lines differing only in their susceptibility to insects may prove useful in determining the chemical basis of insect resistance.

The role of biochemical factors in varietal resistance has been studied in only a few cases (Painter 1958; Beck 1965; and Hsiao 1969); currently several scientists are investigating these factors. Beck (1965) classified biochemical factors involved in varietal resistance in plants into two major categories—physiological inhibitors and nutritional deficiencies.

A classic example of the presence of physiological inhibitors of insect pests in resistant plant hosts is the occurrence in corn inbreds of chemicals toxic to the European corn borer. Two types of plant biochemicals that inhibit the growth of young borer larvae were discovered by Beck and Stauffer (1957), who described them as resistant factor A (ether soluble) and resistant factor B (ether insoluble). Subsequently, an additional resistance factor (C) was found in factor A (Beck 1957). Factors A and C have been isolated and identified as 6-methoxybenzoxazolinone (6-MBOA) and 2,4-dihydroxy-7-methoxy-1,4-benzoxazine-3-one (DIMBOA), respectively. Factor C is a biochemical precursor of factor A. Factor B was found to be of relatively minor importance in borer resistance, and its chemical identity has not yet been established. The concentration of these factors depends on variety and stage of plant growth. The resistance of a corn plant to corn borers depends upon the presence of an effective concentration of the resistant factors in the right tissues at the appropriate stage of growth (Beck 1965).

A strong correlation between the amount of 6-MBOA produced by eleven maize inbreds at the whorl stage of development and the field rating of resistance of inbreds to the first-brood European corn borer was reported by Klun and Brindley (1966). Highly resistant inbreds yielded ten times more 6-MBOA than highly susceptible corn inbreds. DIMBOA was found to be biologically more active than 6-MBOA (Klun, Tipton, and Brindley 1967). The possibility of grading for resistance to the corn borer based on the DIMBOA content of the plant was investigated using eleven inbreds and their diallele progenies (Klun et al. 1970). Leaf-feeding ratings were obtained under artificial infestation and whorl-leaf samples were analyzed for DIMBOA. There was linear correlation between the DIMBOA content in the corn plant and the resistance of corn borers to leaf feeding for both inbred and hybrid corn lines. The higher the DIMBOA level in the plant leaf, the greater was the borer resistance to leaf feeding. The variance for general combining ability accounted for 84% of the variation in resistance ratings and 91% of the variation in concentration in DIMBOA. Also, Reed, Brindley, and Showers (1972) demonstrated that the mortality of the corn-borer larvae feeding on leaves of three selected corn inbreds was linearly related to the concentration of DIMBOA in the leaf tissue.

The resistant factor A (6-MBOA) acts as a growth inhibitor of a variety of organisms, including bacteria, free-living and pathogenic fungi, and a number of insects (Virtanen and Hietala 1955; Whitney and Mortimore 1959 and 1961; Beck and Smissman 1961). Although the implication that certain endogenous plant chemicals can impart resistance to several pests and diseases has not been explored, many plant varieties have been found that show resistance to a wide variety of pests and diseases (Pathak 1972; International Rice Research Institute 1972).

Munakata and Okamoto (1967) reported the presence of benzoic acid and salicylic acid in rice plants and suggested that these acids were toxic to the striped borer larvae. However, the role of these substances in varietal resistance in rice to the striped borer has not been investigated. The presence of benzyl alcohol in barley plants resistant to greenbug, *Schizaphis graminum* (Rondani), has been suggested as a possible cause of this resistance (Juneja et al. 1972). The addition of benzyl alcohol to the hydroponic medium rendered the susceptible variety resistant.

The green peach aphid, *Myzus persicae* (Sulzer), feeds on the phloem of the tobacco plants and thereby avoids a powerful toxin in the xylem tissues (Guthrie, Campbell, and Baron 1962). However, the aphid does not survive on *Nicotiana gossei* because the toxin in this species of tobacco exudes from leaf hairs and kills the aphid by contact effects (Thurston and Webster 1962).

The importance of nutritional deficiency in host plants is not fully understood, although nutritional factors have been implicated in varietal resistance of some plants to a few insect pests. Varieties of peas that were resistant to the pea aphid, *Acyrthosiphon pisum* (Harris), were generally deficient in amino acids and were thus less nutritious than the susceptible pea varieties (Auclair 1963).

The resistance of the rice variety Mudgo to the brown planthopper was suggested by Sogawa and Pathak (1970) to be dependent on the variety's asparagine content, which was lower than that of rice varieties susceptible to the insect. Young females caged on Mudgo plants had underdeveloped ovaries that contained few mature eggs, while those caged on susceptible rice varieties had normal ovaries full of eggs.

BIOTYPES

It is natural to expect insects to develop new biotypes that are capable of surviving on resistant plant varieties as these varieties are planted over large areas. The development of biotypes (or physiological races, as they are properly called) within fungus species that cause plant diseases has been a major factor limiting the use of resistant plant varieties. With some fungi, such as the wheat rust, *Puccinia graminis*, several hundred races of the fungus have been identified. However, biotypes are comparatively rare among insect pests, primarily due to two factors—the insect's own complex physiology and the fact that often the resistance of the host plant to the pest is governed by several factors. For example, the resistance of a plant may be due to antibiosis as well as nonpreference reactions, each of which may be the result of several interdependent factors. In general, biotypes are believed to be less likely to develop on plants with a polygenic type of resistance than on plants that have monogenic resistance.

Painter (1966) classified insect biotypes into two broad categories: (1) large and vigorous biotypes which are able to feed on resistant plants, for example, pea aphids on peas; and (2) biotypes which are related to specific genes for resistance, for example, brown planthopper, Hessian fly, and raspberry aphid. The commonly known cases of insect biotypes related to resistant and susceptible plants are listed in table 2.

Four of the seven insects with known biotypes listed in table 2 are aphids. Since most aphids commonly reproduce parthenogenetically, an individual capable of feeding on a resistant plant can build up a new biotype—somewhat similar to fungi, where each mutant nucleus is capable of forming a new race. A new biotype (biotype C) of greenbug, *Schizaphis graminum*, threatened serious losses to the sorghum crop, but fortunately sources of resistance to this biotype are also known (Wood 1969). The four known biotypes in corn leaf aphid, *Rhopalosiphum maidis*, differ in their survival and fecundity on, and damage to, different hosts; their geographic distribution; and their tolerance to environmental conditions (Cartier and Painter 1956; and Pathak and Painter 1958 *a*, *b*). Each of the two known biotypes in brown planthopper appears to be adapted to a specific gene (table 3).

Varietal resistance to insects has been long lasting in some plants. Wheat varieties grown in Kansas and California have been resistant to the Hessian fly, *Mayetiola destructor*

TABLE 2

Insect biotypes involved in plant resistance (Pathak 1970)

Insect	Common Name	Crop	Biotypes (No.)
Acyrthosiphon pisum	pea aphid	peas, alfalfa	3,9
Amphoraphora rubi	aphid	raspberry	3
Nilaparvata lugens	brown plant-hopper	rice	2
Phytophaga destructor	Hessian fly	wheat	6
Rhopalosiphum maidis	corn leaf aphid	sorghum, corn	4
Schizaphis graminum	greenbug	wheat	3
Therioaphis maculata	spotted alfalfa aphid	alfalfa	1

TABLE 3

Reactions of rice varieties to biotypes of the brown planthopper (Athwal and Pathak 1972)

Variety	Reaction to Brown Planthopper		Genes for Resistance
	Biotype 1	Biotype 2	
Taichung Native 1	S	S	bph 1 bph 1 Bph 2 Bph 2
Mudgo	R	S	Bph 1 Bph 1 Bph 2 Bph 2
CO 22	R	S	
MTU 15	R	S	
MGL 2	R	S	
IR747B2-6	R	S	
ASD 7	R	R	bph 1 bph 1 bph 2 bph 2
PTB 18	R	R	

NOTE: R = resistant; S = susceptible; Bph, bph = brown planthopper.

(Say), since about 1946. In Indiana, although extensive cultivation of varieties containing the H_3 gene for resistance caused a shift from predominance of fly race A to race B, most wheat varieties now grown carry resistance to both these races and provide practical control of the Hessian fly (Hatchett and Gallun 1968). Also, wheat varieties known to be resistant to each of the six biotypes of Hessian fly are available (Gallun, Deay, and Cartwright 1961; Hatchett 1969).

Callenbach (1951) found no evidence of the development of biotypes in wheat stem sawfly, as related to the resistant wheat variety Rescue. Biotypes in European corn borer have not been recorded although some of the resistant wheat inbreds have been continuously grown since 1949. Furthermore, several of these inbreds were also rated resistant to corn-borer populations in Switzerland, Yugoslavia, and Israel (Guthrie 1969).

Since polygenic resistance tends to restrict and reduce the development of biotypes in insects, pooling as many sources of resistance as possible is recommended.

Genetics of Resistance

A knowledge of the genetics of plant resistance to insect pests can aid in the development of varieties which have diverse genetic sources of insect resistance, thus minimizing the risk of the development of insect biotypes adapted to resistant plants. Although this knowledge is useful in developing isogenic resistant plant lines, which are valuable materials for breeding purposes, and in determining causes of plant resistance to pests and diseases, it is not essential in breeding for resistant plant varieties.

The inheritance of insect resistance in plants has been studied only for a few insects. The usual procedure is to test the F_2 segregates and backcross progenies, but diallele crosses are also used where several resistant or susceptible varieties are involved. Evaluation of F_3 lines whereby the resistant ratings are based on pedigree rows rather than on individual plants is preferred in studies of IRRI.

Since physical factors alone may not be the sole cause of resistance, plant resistance to pests may need to be evaluated by exposing them to an infestation of the insect. The completeness of the information on the number of genes involved in insect resistance depends on the thoroughness with which the germ plasm has been screened. For example, varieties coming from one area might have derived their resistance from a common parent, but additional genes of resistance may be available in materials from another locale.

Examples of the information available on the inheritance of varietal resistance of six crop plants to insect pests is presented in table 4.

The inheritance of Hessian fly resistance in wheat has been extensively investigated. At present five dominant or partially dominant genes and five recessive genes are being utilized in breeding wheat that has resistance to the Hessian fly (Abdel-Malik, Heyne, and Painter 1966).

The inheritance of resistance to European corn borer in corn inbreds was studied by Penny and Dicke (1956), who identified three pairs of genes with partial phenotypic dominance of susceptibility in the cross M 14 (S) x MS 1 (R), one or two pairs of genes in B 14 (S) x N 32 (R), and a single dominant gene in WF 9 (S) x gl_7V_{17} (R). Most of the genetic variance of corn to this corn borer is believed by Scott, Hallauer, and Dicke (1964) to be of the additive type, although a portion of the variance was of the dominant type. Inheritance of second-brood resistance is unknown, although data from forty-five diallele crosses among ten inbred lines have shown that the resistance is transmitted in hybrid combinations (Guthrie 1969).

Resistance to the brown planthopper, *Nilaparvata lugens*, and the green leafhopper, *Nephotettix virescens* (Distant), in rice varieties is simply inherited. One dominant (Bph 1) and one recessive (bph 2) gene have been identified in rice for resistance to brown planthopper. Three independently inherited dominant genes (Glh 1, Glh 2, Glh 3) have been identified for resistance to green leafhopper. The two genes for resistance to *N. lugens* are allelic or closely linked, but the three genes for resistance to *N. virescens* are independent of each other (Athwal and Pathak 1972).

In cases of polygenic resistance, recurrent selection or diallele systems of crossing can be used to increase further the existing levels of plant resistance to an insect pest. This is being investigated against the striped borer at IRRI where several progenies have been found which are about four times more resistant than any of the parents used.

Developing Insect-Resistant Plant Varieties

Until the mid-1960s the study of host-plant resistance as a method of insect control received little attention, except in a few cases where other methods of insect control were not practical. Such studies require close collaboration among entomologists, plant breeders, geneticists, and often

TABLE 4

Examples of information on genetics of the resistance of crop plants to insect pests (Pathak 1970)

Crop	Insect Pest	Nature of Resistance	Reference
Corn	European corn borer	Polygenic, additive dominance	Guthrie 1969
	Western corn rootworm	Monogenic, recessive	Sifuentes and Painter 1964
Rice	Striped rice borer	Polygenic	Athwal and Pathak 1972
	Brown plant-hopper	One dominant and one recessive	Athwal and Pathak 1972
	Rice green leafhopper	Monogenic, dominant, 3 different identified genes	Athwal and Pathak 1972
	Rice gall midge	Three pairs of recessive genes	Shastry et al. 1972
Rye	Greenbug	Monogenic, dominant	Livers and Harvey 1969
Sorghum	Shoot fly	Complex, additive, some epistasis	Starks, Eberhart, and Dogget 1969
Sweetclover	Sweetclover aphid	Monogenic, dominant	Manglitz and Gorz 1968
Wheat	Hessian fly	Five dominant or partially dominant and 5 recessive genes	Abdel-Malik, Heyne, and Painter 1966
	Greenbug	Monogenic, recessive with possibility of modifiers	Painter and Peters 1956

biochemists and require several years to develop a resistant plant variety; this approach thus appeared cumbersome when compared with using chemical pesticides to knock down pest populations. However, the growing awareness of the limitations of pesticides in recent years has caused an upsurge of interest in this approach to insect control in crop production.

In a project to develop plant varieties with resistance to a specific insect pest, screening of the available germ plasm of host plants for resistance to that specific pest is the first step. In general, the greater the diversity of types of germ plasm evaluated, the greater are the possibilities for locating resistant lines. Resistance to a particular insect-pest species is frequently, but not always, found in varieties or selections from areas indigenous to the pest. Resistance to the chinch bug, *Blissus leucopterus*, in sorghum varieties from Africa (Painter 1951) and to rice delphacid, *Sogatodes orizicola*, in rice varieties from Southeast Asia (Jennings and Pineda 1970) are examples of natural resistance in plant materials from areas where these insects have not been recorded to occur.

The preliminary screening tests are normally designed to eliminate the bulk of the susceptible varieties or test lines; then the remaining few lines are retested intensively to determine the consistency of their resistance and to discard the pseudoresistant plants, including cases of escapes and host evasions (Painter 1951). Studies on the inheritance of resistance facilitate the identification of different genes for resistance. These genes can also be sought out among other species of host plants and in mutants of the same species.

Once the sources of resistance have been identified, the resistant factor(s) must be incorporated into plants possessing good agronomic qualities. Close collaboration among entomologists and plant breeders, and frequently plant pathologists and agronomists, is needed even though the actual procedures of breeding for insect resistance are simple. Although information on the exact causes of resistance or its genetics is an important addition to a basic understanding of varietal resistance, experience has shown that it is not essential for developing plant varieties resistant to insect pests.

Many crops are subject to infestation by a wide variety of insect pests and may have other problems as well; resistance to a particular pest species is only a part of the overall required characteristics of a plant. Efforts are now under way on several crops to incorporate simultaneously into new plant lines not only resistance to pests and diseases, but also other agronomic characteristics.

The current varietal development program at IRRI aims at incorporating resistance to five different insect pests and seven diseases and tolerance to five different problem soils, drought, flood, deep water, and cold. Close collaboration among specialists is needed to accomplish these goals. In the IRRI program employing interdisciplinary cooperation, a formalized Genetic Evaluation and Utilization program (GEU) (fig. 1) has been established. The procedural details of this program are outlined in figure 2. IRRI has a collection of about thirty thousand different rices from all over the world which are being evaluated for their resistance to various problems. About two thousand crosses are evaluated each year, and this number is expected to increase to three thousand by 1975. Selected collections are being used to investigate the nature of resistance in the breeding program. Additionally, simultaneous evaluation of these materials for various problems frequently leads to identification of rice lines that are multiresistant and can serve as valuable breeding materials. Also, the interdisciplinary-team approach greatly enhances the probability and rate of pooling a large number of desirable characters into a newly developed rice line.

ROLE OF RESISTANT PLANT VARIETIES IN INSECT CONTROL

The antibiosis type of varietal resistance is the primary influence against pest populations, although nonpreference frequently plays a major role. The nonpreferred plant varieties limit insect feeding, thereby producing less vigorous, smaller-sized, and shorter-lived adults. Furthermore, even normal adults from susceptible hosts when caged on nonpreferred plants frequently lay a smaller number of eggs than they do on preferred hosts. The presence of an ovipositional arrestant for boll weevil in certain cotton varieties is an excellent example of this reaction (Maxwell et al. 1969).

Painter (1958) reported that insects that fed on plants possessing an antibiosis type of resistance died or had prolonged nymphal periods, reduced longevity, smaller bodies, and reduced food reserves sometimes followed by unsuccessful hibernation and frequently exhibited restlessness and other behavioral peculiarities. These factors have a cumulative

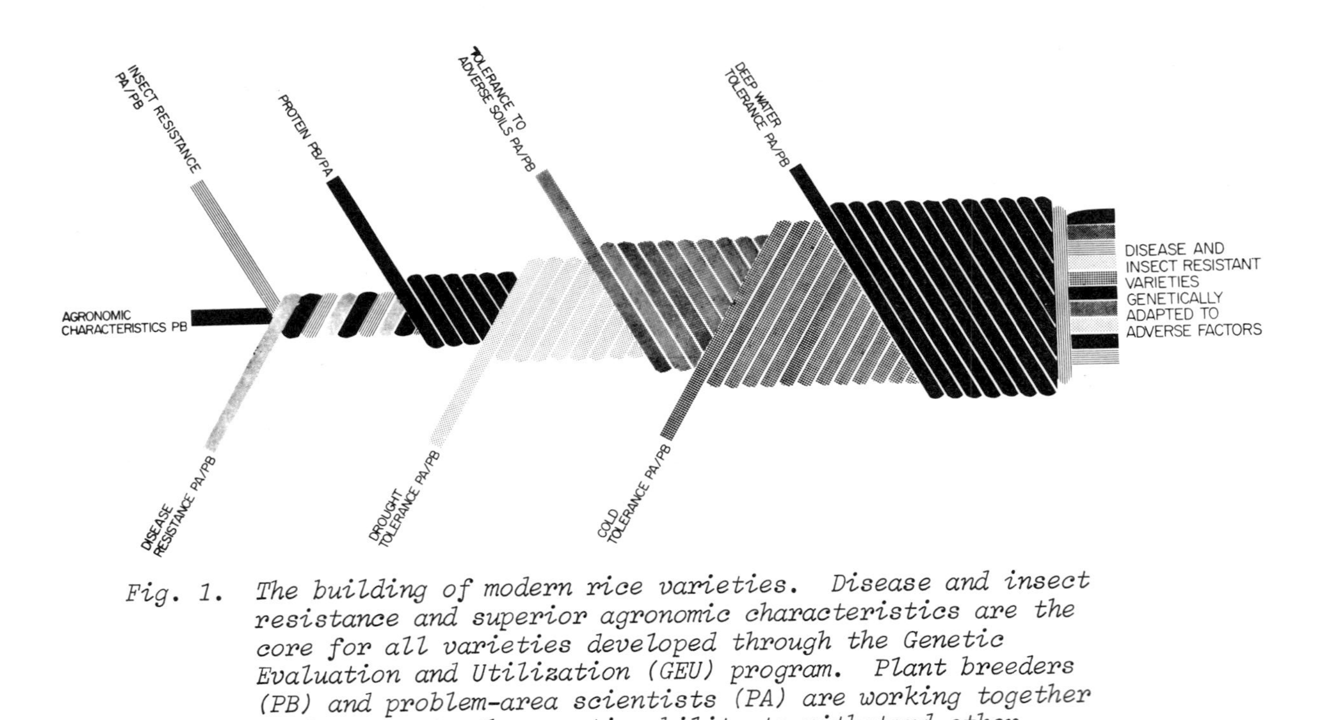

Fig. 1. *The building of modern rice varieties. Disease and insect resistance and superior agronomic characteristics are the core for all varieties developed through the Genetic Evaluation and Utilization (GEU) program. Plant breeders (PB) and problem-area scientists (PA) are working together to incorporate the genetic ability to withstand other production constraints (IRRI 1973).*

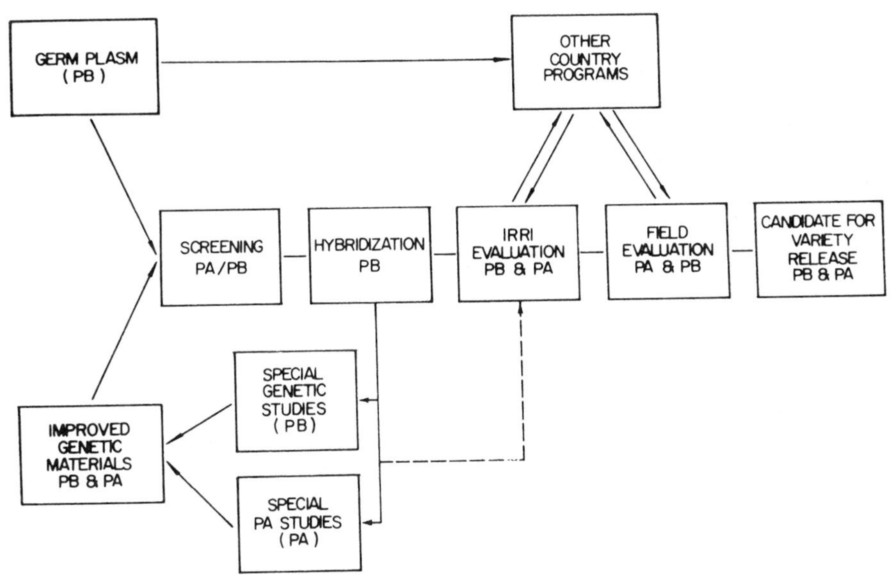

Fig. 2. General outline of GEU (IRRI 1973).

effect which becomes increasingly evident when studied for several generations in cage tests or on large-area plantings of resistant plant varieties.

Plant varieties resistant to more than a dozen insect-pest species have been developed and released for cultivation. Several have been under cultivation for many years, thereby causing marked reductions in pest populations. However, in most cases quantitative data on the effect of long-term cultivation of resistant plant varieties on insect populations are not available.

Dahms (1969), who worked on a theoretical model of greenbug and resistant and susceptible barley varieties, showed that, if the insects reproduced only twice as fast on the susceptible as on the resistant variety and all other factors were constant, at 15 days after caging (the insect completes its life cycle in about one week) its population would be 3 times greater and at 50 days, 40 times greater on the susceptible than on the resistant variety. With this model, considering the fact that the nymphal period is prolonged on resistant varieties, the population at the end of 50 days could be lower by a factor of 1500 on the resistant variety. These differences may actually be even greater, since these calculations do not take into account mortality of nymphs, a very common effect of antibiosis.

Pathak (1970) reviewed information on twenty-one different insect species, and in most of these cases the increase of insect populations on susceptible plants within one generation was five- to tenfold that on resistant plants. Thus it is expected that within a few generations most of these pests will differ greatly in the populations on resistant and susceptible plant hosts.

In nature several other factors adversely influence insect populations; thus the actual population increases would not be at the calculated rates. Nevertheless, such differences in population trends are often maintained on resistant and susceptible hosts.

The actual development of insect populations on resistant and susceptible varieties of several rice-insect pests has been studied at IRRI. In one such experiment on the striped borer, in 120 days after the initial infestation, the borer population increased 30 times more on the susceptible variety Sapan Kawai than on the resistant variety Chianan 2 (Pathak et al. 1971). In a similar experiment with the striped borer using another set of rice varieties, the differences were even larger (fig. 3). Fewer larvae survived on the resistant variety Taitung 16 than on the susceptible variety Rexoro. Their rate of survival on the resistant variety declined in each subsequent generation. On the moderately resistant rice variety IR20, the larval survival in the first generation was only slightly lower than that on the susceptible variety, but it tended to decline in subsequent generations. Furthermore, the larvae reared on resistant or moderately resistant varieties were smaller, took longer to reach pupal stage, and gave rise to the smaller moths which, in turn, laid fewer eggs.

The differences in the rate of population build-up on resistant and susceptible rice plants are more marked in the green leafhopper and the brown planthopper pests. Because the varietal resistance to these insects is high, frequently the insects caged on resistant plants die within the first generation, whereas they multiply rapidly on susceptible rice varieties (Pathak 1970). The population build-up of the white-backed planthopper, *Sogatella furcifera* (Horvath), for which the level of varietal resistance is somewhat lower than for both the green leafhopper and the brown planthopper, is shown in table 5. Within 60 days or approximately three generations after caging, the insect population was 20-30

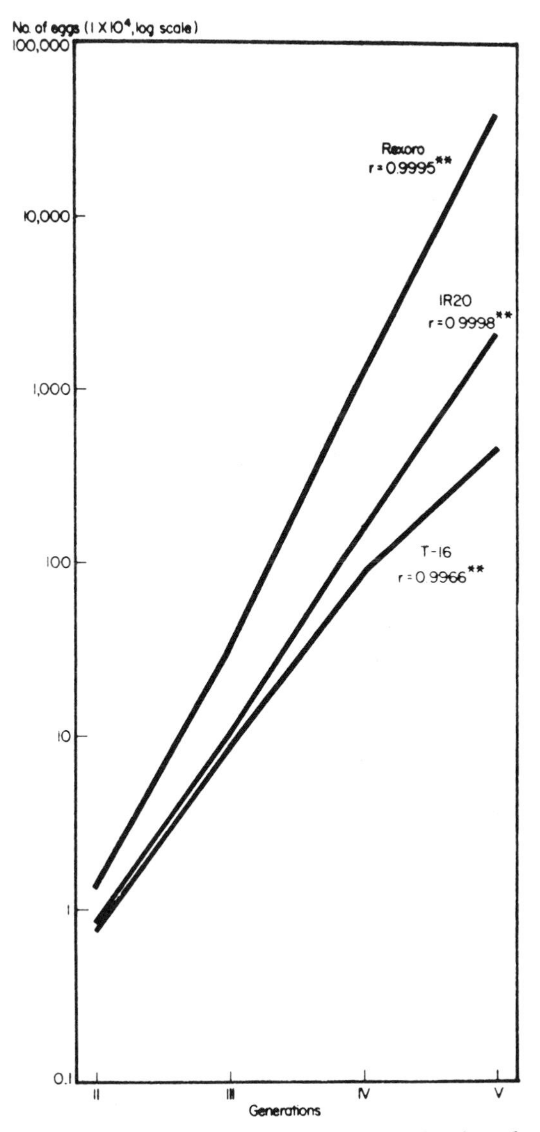

Fig. 3. *Population build-up of striped rice borers on three varieties. An initial population of 350 first-instar larvae was caged on each variety (IRRI 1972).*

TABLE 5

Build-up of whitebacked planthopper populations from an initial caging of five pairs of newly emerged adults on selected rice varieties (IRRI 1972)

Variety	Mean* No. Insects Recovered					
	30 Days after Infestation			60 Days after Infestation		
	Adults	Nymphs	Total	Adults	Nymphs	Total
ARC 5752 (resistant)	35 a	22 a	57 a	16 a	12 a	28 a
Colombo (resistant)	43 a	68 bc	111 b	27 ab	48 b	78 bc
C 5-17 (resistant)	43 a	62 b	113 b	36 b	17 a	48 b
Taichung Native 1 (susceptible)	337 f	154 e	532 g	342 d	6936 c	7316 d

*Values are transformed to log (x + 1).
NOTE: Any two means followed by at least one common letter are not significantly different at the 5% level.

times greater on the susceptible variety than on the resistant rice variety.

The brown planthopper is becoming a major problem in the tropics, apparently due to the changes in the plant type of rice (from tall traditional to modern semidwarf) and the greater use of nitrogenous fertilizers. The population build-up of this insect on resistant and susceptible varieties at different levels of applied nitrogen fertilizer is shown in table 6. On Taichung Native 1 the pest population increased with an increase in the quantity of fertilizer applied, but no such distinct effect was observed in the resistant rice variety Mudgo.

In some areas of Southeast Asia the brown planthopper had at one time become so abundant that it was not possible to grow rice without intensive protection by insecticides. However, excellent crops of resistant varieties are now being grown in these areas without any insecticidal protection. The brown-planthopper population appears to be declining in

TABLE 6

Effect of nitrogen fertilizer on the reaction of two rice varieties to the brown planthopper (IRRI 1972)

Nitrogen (kg/ha)	Insect Survival[a] (%) Taichung Native 1	Insect Survival[a] (%) Mudgo	Male:Female Ratio[b] Taichung Native 1	Male:Female Ratio[b] Mudgo	Insect Progeny Produced[c] (No.) Taichung Native 1	Insect Progeny Produced[c] (No.) Mudgo
0	30	2	1:2.3	1:0.66	4775	11
50	38	0	1:1.4	1:0.71	5139	0
100	44	10	1:1.2	1:0.5	6835	19
150	54	22	1:1.4	1:1	8875	85
200	57	18	1:1.6	1:1.1	9363	70

[a] 22 days after infestation with first-instar nymphs
[b] 17 days after infestation with first-instar nymphs
[c] 37 days after infestation with first-instar nymphs

some of these areas, but long-term data are needed for conclusive evidence. The brown planthopper could be a vicious problem—it takes about 400-500 insects/hill to cause hopperburn and there are about 160,000 hills/ha. Even under these conditions, resistant rice varieties such as IR26 flourish.

Extensive data are available on the effect of resistant plant varieties on the field populations of the Hessian fly and the wheat stem sawfly. At one time both of these insects were serious problems, but their status was reduced to that of minor pests after the cultivation of resistant varieties (Painter 1958; Luginbill 1969; Luginbill and Knipling 1969). Similarly, alfalfa varieties resistant to spotted alfalfa aphid, *Therioaphis maculata* (Buckton), and corn varieties resistant to the European corn borer are contributing significantly to the control of these pests (Luginbill 1969; Chiang 1968; Pesho, Dicke, and Russell 1965; Guthrie 1969).

Varietal Resistance as a Supplement to Other Methods

The unique advantage of varietal resistance is its compatability with other methods of pest control. Insects feeding on resistant hosts are often less vigorous and smaller and thus have greatly reduced resistance to environmental variations. Their restless behavior on resistant plants exposes them more to predators that may devour more small-sized insects from resistant plants than larger insects from susceptible hosts. The bollworm larvae hatching from eggs placed on plant apices remain longer on the terminal growth of Frego resistant cotton than on the susceptible DPL-SL variety (Lincoln et al. 1971). Predators killed about twice as many larvae on the resistant variety as on the susceptible variety.

Furthermore, resistant plant varieties with smaller insect infestations may require fewer insecticidal treatments than do susceptible varieties. Insect-resistant rice varieties receiving only one or two insecticidal treatments had borer infestations identical to those of susceptible varieties receiving twice as many insecticidal treatments (Pathak 1967; IRRI 1968, 1969). With a single application of diazinon, damage by the sweet potato flea beetle, *Chaetocnema confinis* (Crotch), to the roots of resistant sweet potato variety NC240 was 3%, while a 43% damage was sustained by the susceptible variety Julian (Brett and Sullivan 1969). Chalfant and Brett (1967) recorded that at moderate infestations of the cabbage looper, *Trichoplusia ni* (Hübner), and the imported cabbage worm, *Pieris rapae* (Linnaeus), resistant cabbage varieties responded more favorably to insecticidal treatments than did the susceptible varieties.

Resistance to Vectors to Reduce Virus Incidence

Phytoresistance to virus has not been generally correlated with resistance to vectors. However, the concept that vector resistance in plants can minimize considerably the spread of virus diseases has not been widely appreciated. Some significant examples of this are available in rice.

Although the rice variety IR20 is susceptible to the *tungro* virus in greenhouse inoculation tests, it is resistant to the green leafhopper *Nephotettix virescens* (Horvath), which serves as the vector of the *tungro* virus. Thus in field plantings IR20 rarely is infected with *tungro* virus, even when the virus is present in epidemic proportions.

Rice variety IR8 is resistant to the rice delphacid in Colombia but is susceptible to *hoja blanca* transmitted by this insect. In field plantings IR8 remains virtually virus free, apparently because of its resistance to the vector, while other susceptible rice lines in adjacent plots become heavily infected (Jennings and Pineda 1970).

Similarly, many rice varieties resistant to the brown planthopper, but susceptible to grassy stunt virus transmitted by it, have consistently demonstrated field resistance to this virus. For reasons still unexplained, however, rice plants which are resistant to the vectors of *tungro* virus and grassy stunt virus differ in their susceptibility to both of these virus diseases.

CONCLUSION

The field of varietal resistance of crop plants to insect infestations offers a tremendous practical method of insect control. Interdisciplinary cooperation in this area of research is needed to increase our understanding of the nature of plant resistance to pests and to develop insect-resistant varieties of crop plants.

REFERENCES

Abdel-Malik, S.H., E.G. Heyne, and R.H. Painter. 1966. Resistance of F_1 wheat plants to green bugs and Hessian fly. J. Econ. Entomol. 59:707-10.

Athwal, D.S., and M.D. Pathak. 1972. Genetics of resistance to rice insects, p. 375-86. *In* Int. Rice Res. Inst., Rice breeding. Los Baños, Philippines.

Auclair, J.L. 1963. Aphid feeding and nutrition. Annu. Rev. Entomol. 8:439-90.

Beck, S.D. 1957. The European corn borer, *Pyrausta nubilalis* (Hübner), and its principal host plant. VI. Host plant resistance to larval establishment. J. Insect Physiol. 1:158-77.

———. 1965. Resistance of plants to insects. Annu. Rev. Entomol. 10:207-32.

Beck, S.D., and E.E. Smissman. 1961. The European corn borer, *Pyrausta nubilalis*, and its principal host plant. IX. Biological activity of chemical analogs of corn resistance factor A (6-methoxybenzoxazolinone). Ann. Entomol. Soc. Amer. 54:53-81.

Beck, S.D., and J.F. Stauffer. 1957. The European corn borer, *Pyrausta nubilalis* (Hübner), and its principal host plants. III. Toxic factor influencing larval establishment. Ann. Entomol. Soc. Amer. 50:166-70.

Brett, C.H., and M.J. Sullivan. 1969. Sweet potato flea beetle control. Vegetable Insect Annual Report, N.C. State University, Raleigh.

Callahan, P.S. 1957. Oviposition response of the imago of corn earworm, *Heliothis zea* (Boddie), to various wave lengths of light. Ann. Entomol. Soc. Amer. 50:444-52.

Callenbach, J.A. 1951. Rescue wheat and its resistance to wheat stem sawfly attack. J. Econ. Entomol. 44:999-1001.

Cartier, J.J., and R.H. Painter. 1956. Differential reactions of two biotypes of the corn leaf aphid to resistant and susceptible varieties, hybrids and selections of sorghums. J. Econ. Entomol. 49:498-508.

Chalfant, R.B., and C.H. Brett. 1967. Interrelationship of cabbage variety, season, and insecticide, on control of the cabbage looper and the imported cabbage worm. J. Econ. Entomol. 60:687-90.

Chiang, H.C. 1968. Host variety as an ecological environmental factor in the population dynamics of the European corn borer, *Ostrinia nubilalis*. Ann. Entomol. Soc. Amer. 61:1521-23.

Dahms, R.G. 1969. Theoretical effects of antibiosis on insect population dynamics. U.S. Dep. Agr., Econ. Res. Serv., Beltsville, Maryland.

Djamin, A., and M.D. Pathak. 1968. Role of silica in resistance to Asiatic borer, *Chilo suppressalis* (Walker), in the rice varieties. J. Econ. Entomol. 60:347-51.

Doane, J.F., and R.K. Chapman. 1962. Oviposition preference of the cabbage maggot, *Hylemya brassicae* (Bouche). J. Econ. Entomol. 55:137-38.

Gallun, R.L., H.O. Deay, and W.B. Cartwright. 1961. Four races of Hessian fly selected and developed from an Indiana population. Purdue Univ. Res. Bull. 732.

Guthrie, W.D. 1969. European corn borer investigations. European Corn Borer Laboratory, Ankeny, Iowa.

Guthrie, F.E., W.V. Campbell, and R.L. Baron. 1962. Feeding sites of the green peach aphid with respect to its adaptation to tobacco. Ann. Entomol. Soc. Amer. 55:42-46.

Guthrie, W.D., and E.V. Walter. 1961. Corn ear worm and European corn borer resistance in sweet corn inbred lines. J. Econ. Entomol. 54:1248-50.

Hatchett, J.H. 1969. Race E, sixth race of the Hessian fly, *Mayetiola destructor*, discovered in Georgia wheat fields. Ann. Entomol. Soc. Amer. 62:677-78.

Hatchett, J.H., and R.L. Gallun. 1968. Frequency of Hessian fly, *Mayetiola destructor*, races in field populations. Ann. Entomol. Soc. Amer. 61:1446-49.

Howe, W.L. 1949. Factors affecting the resistance of certain cucurbits to the squash borer. J. Econ. Entomol. 42:321-26.

Hsiao, T.H. 1969. Chemical basis of host selection and plant resistance in oligophagous insects, p. 777-88. *In* Proc. Second Int. Symp. on Insect and Host Plant, Wageningen.

International Rice Research Institute. 1968. Annual report for 1967. Los Baños, Philippines.

———. 1969. Annual report for 1968. Los Baños, Philippines.

———. 1972. Annual report for 1971. Los Baños, Philippines.

———. 1973. Research highlights for 1973. Los Baños, Philippines.

Jennings, P.R., and A. Pineda. 1970. *Sogatodes orizicola* resistance in rice varieties. Centro Internacional Agricultura Tropical, Palmira, Colombia.

Juneja, P.S., R.K. Gholson, R.L. Burton, and K.J. Starks. 1972. The chemical basis for greenbug resistance in small grains. l. Benzyl alcohol as a possible resistance factor. Ann. Entomol. Soc. Amer. 65:961-64.

Keller, J.C., F.G. Maxwell, and J.N. Jenkins. 1962. Cotton extracts as arrestants and feeding stimulants for the boll weevil. J. Econ. Entomol. 55:800-801.

Keller, J.C., F.G. Maxwell, J.N. Jenkins, and T.B. Davich. 1963. A boll weevil attractant from cotton. J. Econ. Entomol. 56:110-11.

Klun, J.A., and T.A. Brindley. 1966. Role of 6-methoxy-benzoxazolinone in inbred resistance of host plant (maize) to first-brood larvae of European corn borer. J. Econ. Entomol. 59:711-18.

Klun, J.A., W.D. Guthrie, A.R. Hallauer, and W.A. Russell. 1970. Genetic nature of the concentration of 2,4-dihydroxy-7-methoxy *2H*-1,4-benzoxazin-3(*4H*)-one and resistance to the European corn borer in a diallele set of eleven maize inbreds. Crop Sci. 10:87-90.

Klun, J.A., C.L. Tipton, and T.A. Brindley. 1967. 2,4-Dihydroxy-7-methoxy-1,4-benzoxazine-3-one (DIMBOA) an active agent in the resistance of maize to the European corn borer. J. Econ. Entomol. 60:1529-33.

Lincoln, C., G. Dean, B.A. Waddle, W.C. Yearian, J.R. Phillips, and L. Roberts. 1971. Resistance of Frego-type cotton to boll weevil and bollworm. J. Econ. Entomol. 64:1326-27.

Livers, R.W., and T.L. Harvey. 1969. Greenbug resistance in rye. J. Econ. Entomol. 62:1368-70.

Luginbill, P. 1969. Developing resistant plants—the ideal method of controlling insects. U.S. Dep. Agr., Agr. Res. Serv., Prod. Res. Rep. 111.

Luginbill, P., and E.F. Knipling. 1969. Suppression of wheat stem sawfly with resistant wheat. U.S. Dep. Agr., Agr. Res. Serv., Prod. Res. Rep. 107.

Manglitz, G.R., and H.J. Gorz. 1968. Inheritance of resistance in sweetclover to the sweetclover aphid. J. Econ. Entomol. 61:90-93.

Maxwell, F.G., J.N. Jenkins, and J.C. Keller. 1963. A boll weevil repellent from the volatile substances of cotton. J. Econ. Entomol. 56:894-95.

Maxwell, F.G., J.N. Jenkins, and W.L. Parrott. 1972. Resistance of plants to insects. Advan. Agron. 24:187-265.

Maxwell, F.G., J.N. Jenkins, W.L. Parrott, and W.T. Buford. 1969. Factors contributing to resistance and susceptibility of cotton and other hosts to the boll weevil, *Anthonomus grandis*, p. 801-10. *In* Proc. Second Int. Symp. on Insect and Host Plant, Wageningen.

Moore, R.H. 1928. Odorous constituents of the corn plant in their relation to European corn borer. Proc. Oklahoma Acad. Sci. 8:16-18.

Munakata, K., and D. Okamoto. 1967. Varietal resistance to rice stem borer in Japan, p. 419-30. *In* The major insect pests of the rice plant. Proc. Symp. Int. Rice Res. Inst., Los Baños, 1964. Johns Hopkins Press, Baltimore.

Oatman, E.R. 1959. Host range studies of the melon leaf miner *Liriomyza pictella* (Thompson). Ann. Entomol. Soc. Amer. 52:739-41.

Painter, R.H. 1951. Insect resistance in crop plants. MacMillan, New York.

———. 1958. Resistance of plants to insect. Annu. Rev. Entomol. 3:267-90.

———. 1966. Lessons to be learned from past experience in breeding plants for insect resistance, p. 349-66. *In* Breeding pest resistant trees. Proc. North Atlantic Treaty Organ. and Nat. Sci. Found. Symp. Pergamon Press, New York.

Painter, R.H., and D.C. Peters. 1956. Screening wheat varieties and hybrids for resistance to the green bug. J. Econ. Entomol. 49:546-48.

Patanakamjorn, S., and M.D. Pathak. 1967. Varietal resistance of rice to the Asiatic rice borer, *Chilo suppressalis* (Lepidoptera: Crambidae), and its associations with various plant characters. Ann. Entomol. Soc. Amer. 60:287-92.

Pathak, M.D. 1967. Varietal resistance to rice stem borers at IRRI, p. 405-18. *In* The major insect pests of the rice plant. Proc. Symp. Int. Rice Res. Inst., Los Baños, 1964. Johns Hopkins Press, Baltimore.

———. 1970. Genetics of plants in pest management, p. 138-57. *In* R.L. Rabb and F.E. Guthrie [eds.], Concepts of pest management. N.C. State University, Raleigh.

———. 1972. Resistance to insect pests in rice varieties, p. 325-41. *In* Int. Rice Res. Inst., Rice Breeding. Los Baños, Philippines.

Pathak, M.D., F. Andres, N. Galacgac, and R. Raros. 1971. Resistance of rice varieties to striped rice borers. Int. Rice Res. Inst. Tech. Bull. 11.

Pathak, M.D., and R.H. Painter. 1958*a*. Differential amounts of materials taken up by four biotypes of corn leaf aphids from resistant and susceptible sorghums. Ann. Entomol. Soc. Amer. 51:250-54.

———. 1958*b*. Effect of the feeding of the four biotypes of corn leaf aphid, *Rhopalosiphum maidis* (Fitch), on susceptible White Martin sorghum and Spartan barley plants. J. Kans. Entomol. Soc. 31:93-100.

Penny, L.H., and F.F. Dicke. 1956. Inheritance of resistance in corn to leaf feeding by the European corn borer. Agron. J. 48:200-204.

Pesho, G.R., F.F. Dicke, and W.A. Russell. 1965. Resistance of inbred lines of corn (*Zea mays*) to the second brood of the European corn borer (*Ostrinia nubilalis* (Hübner)). Iowa State J. Sci. 40:85-98.

Radcliffe, E.B., and R.K. Chapman. 1960. Insect resistance in commercial varieties of cabbage. Proc. N. Cent. Br. Entomol. Soc. Amer. 15:111-12.

Reed, G.L., T.A. Brindley, and W.B. Showers. 1972. Influence of resistant corn leaf tissue on the biology of the European corn borer. Ann. Entomol. Soc. Amer. 65:658-62.

Ringlund, K., and E.H. Everson. 1968. Leaf pubescence in common wheat, *Triticum aestivum* L., and resistance to the cereal leaf beetle, *Oulema melanopus* (L.). Crop Sci. 8:705-10.

Sasamoto, K. 1961. Resistance of the rice plant supplied with silicate and nitrogenous fertilizers to the rice stem borer, *Chilo suppressalis* Walker. Proc. Fac. Liberal Arts and Educ. 3, Yamanashi Univ., Japan.

Schillinger, J.A. 1969. Three laboratory techniques for screening small grains for resistance to the cereal leaf beetle. J. Econ. Entomol. 62:360-63.

Schillinger, J.A., and R.L. Gallun. 1968. Leaf pubescence of wheat as a deterrent to the cereal leaf beetle, *Oulema melanopus*. Ann. Entomol. Soc. Amer. 61:903.

Scott, G.E., A.R. Hallauer, and F.F. Dicke. 1964. Types of gene action conditioning resistance to European corn borer leaf feeding. Crop Sci. 4:603-4.

Shastry, S.V.S., W.H. Freeman, D.V. Sheshu, P. Israel, and

J.K. Roy. 1972. Host-plant resistance to rice gall midge, p. 353-65. *In* Int. Rice Res. Inst., Rice breeding. Los Baños, Philippines.

Sifuentes, J., and R.H. Painter. 1964. Inheritance of resistance to the western corn rootworm adults in field corn. J. Econ. Entomol. 57:475-77.

Sogawa, K., and M.D. Pathak. 1970. Mechanisms of brown planthopper resistance in Mudgo variety of rice (Hemiptera: Delphacidae). Appl. Entomol. Zool. 5:145-58.

Starks, K.J., S.A. Eberhart, and H. Dogget. 1969. Recovery from shootfly attack in a sorghum diallele. U.S. Dep. Agr., Econ. Res. Serv., Beltsville, Maryland.

Thorsteinson, A.J. 1960. Host selection in phytophagous insects. Annu. Rev. Entomol. 5:193-218.

Thurston, R., and J.S. Webster. 1962. Toxicity of *Nicotiana gossei* Domin. to *Myzus persicae* (Sulzer). Entomol. Exp. Appl. 5:233-38.

Virtanen, A.I.,and P.K. Hietala. 1955. 2(3)-Benzoxazolinone, an antifusarium factor in rye seedlings. Acta Chem. Scand. 9:1543-44.

Wallace, L.E., F.H. McNeal, and M.A. Berg. 1973. Minimum stem solidness required in wheat for resistance to the wheat stem sawfly. J. Econ. Entomol. 66:1121-23.

Whitney, N.J., and C.G. Mortimore. 1959. Isolation of the antifungal substance, 6-methoxybenzoxazolinone, from field corn (*Zea mays* L.) Nature 184:1320.

———. 1961. Effect of 6-methoxybenzoxazolinone on the growth of *Xanthomonas stewartii* and its presence in sweet corn (*Zea mays* var. *saccharata* Bailey). Nature 189:596-97.

Wilde, J. de. 1958. Host plant selection in the Colorado beetle larva (*Leptinotarsa decemlineata* Say). Entomol. Exp. Appl. 1:14-22.

Wood, E.A. 1969. Non-preference, fecundity, and longevity of three greenbug biotypes cultured on greenbug tolerant sorghum species. U.S. Dep. Agr., Spec. Rep. W. 303.

PART IV

Insect Population Dynamics

THE DYNAMICS OF INSECT POPULATIONS

T.R.E. Southwood

STABILITY AND CHANGE

When entomologists first considered the dynamics of populations, two contrasting features were immediately apparent. On one hand there was surprising stability, on the other there were remarkable changes, numbers rising suddenly and falling dramatically.

A classic example of stability is provided by British butterflies; those species that were common over a hundred years ago remain common, those that were scarce are mostly still scarce. A more numerical example, but again based on comparisons between species, is provided by the Heteroptera collected in a light trap at Rothamsted; although 95 species were caught, in the nine seasons of trapping (spread over 23 years, 1933-1956), three were always amongst the six most abundant (Southwood 1960, 1968). Stability in absolute population size is well illustrated by Ehrlich and Gilbert's (1973) study of the tropical rain-forest butterfly *Heliconius ethilla* in northern Trinidad. They estimated the size of a colony over 27 generations; the number of adults varied from 98 to 199 with a mean of 156, but none of the 20-day-interval, successive population estimates was significantly different from the next.

In contrast, other entomologists, particularly those concerned with economic entomology, have been impressed by the changes in numbers. Outbreaks of pests occur in some years, but not in others. Locusts provide an outstanding example of spectacular periodic epidemics (Waloff 1966); there are a number of references to pest outbreaks in records of the ancient world (Southwood 1968).

In the early development of ecological theory some workers concentrated on the phenomenon of stability that is determined by density-dependent processes (i.e., processes acting on a higher proportion of the population as density increases). A.J. Nicholson was a powerful proponent of this view, and he was supported by C. Elton, D. Lack, M. Solomon, and many others.

A most forceful expression of the contrary view was by H.G. Andrewartha and L.C. Birch in their classic *The Distribution and Abundance of Animals* (1954). They emphasized that there was no need to attach special importance to density-dependent factors. Essentially, they viewed populations as chaotic, simply limited by the shortage of time in which conditions were favorable and the rate of increase was positive and, less frequently, by an absolute or relative shortage of resources.

The dispute probably reached its apogee in June 1957 at the Cold Spring Harbor Symposium on "Population Studies: Animal Ecology and Demography." Here A. Milne elaborated his theory, blending the two approaches and stressing that only intraspecific competition was "perfectly density dependent." Populations could, he supposed, fluctuate for long periods at a lower density because of the combined action of density-independent factors and natural enemies (imperfect density-dependent factors). In the early 1960s Chitty and Pimentel both put forward independent theories that highlighted the importance of changes in quality, genetic composition, and birth rate.

After about 1960 ecologists increasingly realized that the two viewpoints were not mutually exclusive (Richards 1961; Bakker 1964; Huffaker and Messenger 1964; Richards and Southwood 1968). It was recognized that each of these theories had added a new dimension to our picture of population dynamics. Additional data have become available over the last twenty years and, as more sophisticated techniques of theoretical and specific analysis are now adopted, one must hope that the study of population dynamics has now moved from a "single factor" to a "synoptic" stage of development (Southwood 1968). As will be shown in this review, stability and change can be fitted within one conceptual framework, and furthermore change, although normally a response to environmental fluctuations, may be produced in a system governed by density dependence.

The term *stability* poses problems of interpretation. Lewontin (1969), in a prescient paper, considered that ecologists would require the general abstract framework provided by the concept of the vector field in n-dimensional space; he defined a variety of dynamical-space concepts, including neighborhood and global stability and relative and structural stability. The biologists' problems in the measurement of stability were emphasized by Watt (1969). There are a number of possible criteria: fluctuations about a mean, changes in the mean with time, the degree of synchrony throughout the range, or the extent of gradients in these measures from one part of the range to another. Extending the essentially biological viewpoint, Orians (1974) considers there are six meanings of the term *stability*: constancy (lack of change), inertia (resistance to perturbations), elasticity (speed of return after disturbance), amplitude (the zone from which this return will occur), cyclic stability (oscillations), and trajectory stability (e.g., succession).

Such classifications are useful; they essentially highlight the components and forms of stability. But I would agree with May (1973) who, in an opening chapter that calls on both theoretical and field insights, concludes, "It intuitively seems sensible to refer to those systems . . . with relatively small fluctuations as 'stable'."

In studying population change biologists have long recognized three pathways of change: mortality, natality, and migration. (The role of the former has been assessed with much greater frequency [and accuracy] than the other pathways; there are still many difficulties for the quantitative determination of the role of migration.)

The action of factors, using these pathways, may be either—

1. regulating, acting in a positively density-dependent manner (killing a higher proportion with a rise in density) and so tending towards the stabilization of numbers; or
2. disturbing, acting in a manner to cause fluctuations in the population, to move it away from its stable level.

These two effects can be compounded, as for example in a regulating factor whose action is very imprecise (Southwood 1967).

Regulating factors may be classified in a variety of ways: the causal agents (predators, intraspecific competition) or time scale (delayed, direct). Walker (1967) has suggested an interesting approach, recognizing pathological factors (those that involve mortality) and behavioral factors (those that involve changes in behavior, avoiding the death of individuals).

From this general background we can now turn to recent theoretical and field studies to determine how a more synoptic, multidimensional view of insect population dynamics is emerging.

THEORETICAL STABILITY ANALYSIS

Our understanding of the possible mechanisms that will contribute to stability in populations has recently been enlarged by a number of analytical studies. There are two principal types of interactions that may be stabilizing; that is, they have the potential of acting in a density-dependent manner:

1. Competition between members of the same species in relation to some resource. In its best-known form this is expressed as the logistic or Verhulst-Pearl equation:

$$\frac{dN}{dt} = rN\,\frac{K - N}{K}$$

   where N = population density, r = intrinsic rate of natural increase, and K = carrying capacity.

2. The interaction between a prey and its natural enemy. Although pioneered by Howard and Fiske and by H. Smith, the formulation of a quantitative theory basic to recent studies depended on the work of Nicholson, especially Nicholson and Bailey (1935), who expressed the numbers of the host (H) in generation n+1 as:

$$H_{n+1} = FH_n \exp(-aP_n)$$

   where F = host's power of increase, a = searching efficiency of the parasite and P_n = parasite population in generation n.

In addition, there are other interactions—for example, interspecific competition and mutualism—which, although related to density, are not normally directly stabilizing. The general development of theories in these areas has been reviewed by Varley, Gradwell, and Hassell (1973) and Williamson (1972). As they point out, observations in the real world have shown that neither competition nor predation always acts in a stabilizing manner, and recent work has illuminated more precisely why these interactions are not always stabilizing.

Single-Species Dynamics

The simple logistic equation is applicable to overlapping generations and represents a smooth approach to the equilibrium population ($N^* = K$). However, in many, if not most, temperate insects, generations are discrete, and a better representation of population change is given by a difference-equation model (May et al. 1974). The simple case is:

$$N_{t+1} = (\lambda N_t^{-b}) N_t \qquad (1)$$

where N_t and N_{t+1} are population size in successive generations, λ = finite net rate of increase ($=e^r$) and b = density-dependent moderator.

Unlike the logistic, this equation has a finite time delay equivalent to the generation time, and because of this it permits the realistic possibility of overshooting. Now the response of the system will be such that, when the population is perturbed from equilibrium, it returns to this value (N^*) during a characteristic return time T_R, as shown in figure 1*a*. Engineers are familiar with this concept of return time in a system with negative feedback. Its magnitude depends on the parameters governing the feedback progress; it is a property of the system.

If an insect population is considered, at time t+1 the next generation will arise; then and only then (in our model) can account be taken of the population size (through the parameters λ and especially b). Now if the return time (T_R) is more than the generation time (τ), then the population will, conceptually, still be approaching the equilibrium at the end of the first generation, and it will continue to do so in a number of decreasingly steep steps (fig. 1*b*). Alternatively, if T_R is much less than the generation time, then the population will overshoot and oscillate around the equilibrium population (fig. 1*c*).

May et al. (1974) show that the characteristic return time for a system described by equation (1) is given by:

$$T_R = \frac{1}{b}$$

Thus the inverse of the density-dependent moderator represents the characteristic return rate. This peculiar relationship, in which a time is equated with a dimensionless number, only holds when generation time is taken as 1. As can be seen graphically from figure 1, when $0 < b < 1$ the population will stabilize, damping monotonically (exponentially) as it approaches the equilibrium; for the condition $1 < b < 2$ oscillatory damping, with overshooting, will occur.

May et al. (1974) have also shown this mathematically and

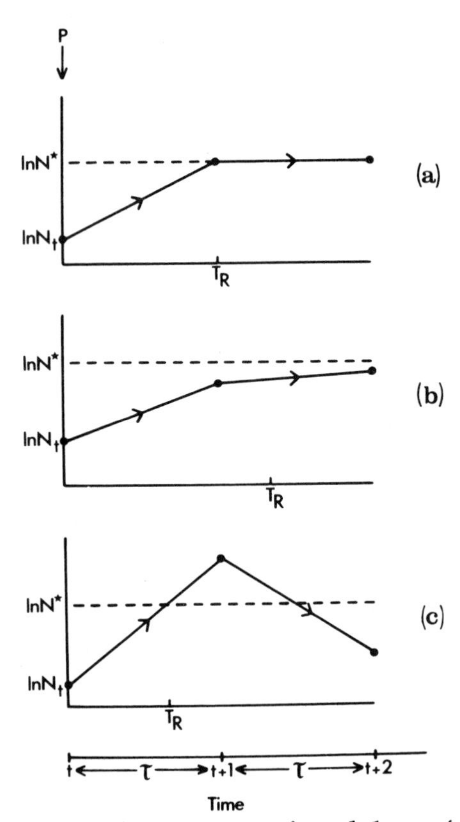

Fig. 1. The relationship between time delays (generation time, τ*) and the characteristic return time of the system* (T_R)*, after perturbation* (P). *(a)* b = *1, return to equilibrium in one generation; (b)* b = *0.7, monotonic return to equilibrium; (c)* b = *1.4, stable limit cycles (oscillations) around equilibrium.*

demonstrated that the same stability criteria apply to the general case with overlapping generations, with a time delay (T) that may depend on the recovery time of a resource as well as generation time, as in the simple example with discrete generations where generation time is taken as unity. This work therefore confirms as a general property what has long been recognized (Haldane 1953; Varley 1963; Southwood 1966; Varley, Gradwell, and Hassell 1973); namely, that values of b in excess of 1 are overcompensating and would lead, at least initially, to oscillations.

Thus one sees that even with a simple model the very parameters that may lead to stability can initially cause change, in the form of oscillations. But real populations are not so simple; insect adults and larvae are often exposed to different ecological factors. May et al. (1974) also investigated the stability properties of such a population; that is, one with two age classes, larvae and adults. Four different b's may be recognized:

b_{LA}—effect on larval recruitment at time t+1 of adult density at time t,

b_{LL}—effect on larval recruitment at time t+1 of larval density at time t,

b_{AL}—effect on adult recruitment at time t+1 of larval density at time t,

b_{AA}—effect on adult recruitment at time t+1 of adult density at time t.

b_{LA} and b_{LL} will act through fecundity, and b_{AL} and b_{AA} through larval survival to adulthood. All of these coefficients interact. For example, if b_{AA} and b_{LL} are zero, then, provided one of the remaining coefficients is around 1, considerable variation is permitted in the other; but if b_{AA} = 1.5, very restricted values of b_{AL} and b_{LA} are necessary for stability.

Therefore natural complexity would enhance the probability of oscillations, unless during natural selection density-dependent moderators of very limited dimensions have been selected. These conclusions are based on the consideration of b's in the range of $0 < b < 2$. What will be the effect of b's of even greater overcompensation, that is, in excess of 2?

These higher values can easily be shown theoretically to lead to diverging oscillations (Haldane 1953; Maynard Smith 1974; May et al. 1974). Their practical significance was doubted by Klomp (1966*a*), but Southwood (1967) suggested that a combination of more sensitive (highly overcompensating) density-dependence mechanisms could be associated with high rates of reproduction and might "show two or four generation cycles."

Most recently, May (in press) has thoroughly investigated the dynamical structure of the nonlinear difference equation:

$$N_{t+1} = N_t \exp[r(1 - N_t/K)] \tag{2}$$

and related formulations (a type of equation relatively neglected by mathematicians). He has considered the effects of the parameter r, which is itself the resultant of two biologically meaningful rates: the gain rate less the loss rate.

The following conclusions could be drawn for the difference equation above:

Value of r	Dynamical Behavior
$2 > r > 0$	stable equilibrium
$2.526 > r > 2$	stable 2-point cycle
$2.692 > r > 2.526$	stable 4-point cycle giving way in turn to 8, 16, 32, etc., as r increases
$r > 2.692$	cycles of arbitrary period or a periodic behavior depending on the initial condition—"chaos"

May (in press) writes that "for temperate-zone insects in particular the implication is that even if the natural world was 100 percent predictable, the dynamics of populations with 'density dependent' regulation could nonetheless in some circumstances be indistinguishable from chaos if the intrinsic growth rate r, is large enough!"

Rates of increase of this magnitude may not be impossibly high. (It should be noted that the density-dependent moderation in this formulation [eq. 2] is provided by the exponent of the term $[1 - N_t/K]$.) An approximate value of r, termed r_c (Laughlin 1965), is given by:

$$r_c = \frac{ln\ R_o}{T_c}$$

where R_o = net reproductive rate and T_c = mean generation time. For insects, where the variance in the mean age of reproduction is small, this approximation will not introduce significant errors. In May's analysis the time interval can be taken as the time from one generation to the next; hence $r \sim ln\ R_o$. The mean fecundities necessary for the critical r's of May may be calculated, if a 50:50 sex ratio is assumed and an arbitrary value is assigned to the density-independent mortality. Klomp (1966*a*) assumed 60%. Although some life tables would appear to support this, higher values may be more realistic. (The deterministic nature of this assumption is a further artificial element.) The values arrived at, using three levels of density-independent mortality, are:

r	$R_o (=e^r)$	Mean Fecundity at 60%	at 80%	at 90%
2	7.39	37	74	148
2.526	12.55	63	126	251
2.692	14.74	74	148	295

A graphical representation of this stability analysis is shown in figure 2; note the decreasing narrowness of the

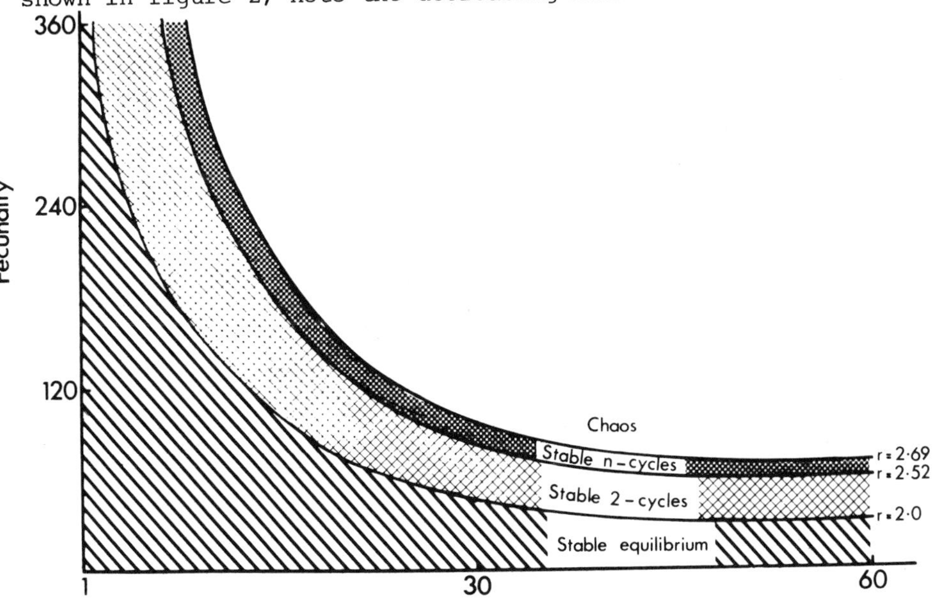

Fig. 2. Stability boundaries for mean fecundity and various levels of density-dependent mortality, based on May's (in press) analysis of r *in the equation* $N_{t+1} = N_t exp[r(1 - N/K)]$.

stable limit cycle bands.

We can conclude from these stability analyses that with insects with discrete generations, where the time delay in the density-dependent feedback may be greater than the natural return time or period of the system (T_R), large oscillations will result, and, where the species can maintain a high fecundity rate at all densities, 2- or more-point stable limit cycles may give way to irregular behavior that will depend on the initial conditions. The interactions of density-dependent coefficients for different stages in a multiple age-class system also theoretically increase the probability of oscillations and fluctuations if the coefficients are randomly selected. Thus it is no longer valid to consider all interactions of a certain type as density dependent, nor to regard all density-dependent relationships as leading to stable populations. The application of these theoretical conclusions to real populations will be considered later.

Predator-Prey Dynamics

Insect parasitoids are, in many respects, ecologically similar to predators and will be included here. However, this topic will only be briefly discussed because it receives separate treatment by Messenger elsewhere in this volume.

Quantitative models start with the work of Thompson and of Nicholson and Bailey; both of those models predicted unstable systems (Huffaker et al. 1968; Royama 1971; Hassell and Rogers 1972) and, although they may be stabilized by the introduction of density dependence into either population, their assumption of a random constant search was biologically unsound (Hassell 1971*b*; Rogers 1972).

Important concepts were introduced by Holling (1959, 1961, 1964, 1966) and Watt (1959), who separated many of the components of the system. In particular they followed Solomon (1949) in distinguishing between two types of response to increases in prey density, the functional response (change in the number of attacks per predator) and the numerical response (change in the number of predators); the latter is now confined to intergenerational effects due to predator reproduction (Hassell 1966). Holling particularly studied the functional-response curve of predators and noted that, although of various forms (the biological universality of these forms has been doubted by Fransz [1974]), the proportion of prey killed would normally fall with increased prey density (fig. 3*a*), and thus this could not provide stability.

Studies have revealed a number of components in a predator-prey system that may contribute to stability. The more important of these are:

1. An individual behavioral response when the predator alters its behavior in the general proximity of the prey (Hassell 1966; Rogers 1972; Murdie and Hassell 1973; Rogers and Hassell 1974; Murdoch and Oaten, in press). This may involve attraction, a response to the changed behavior of another predator that has perceived a prey (as in vultures), or increased turning. These lead to

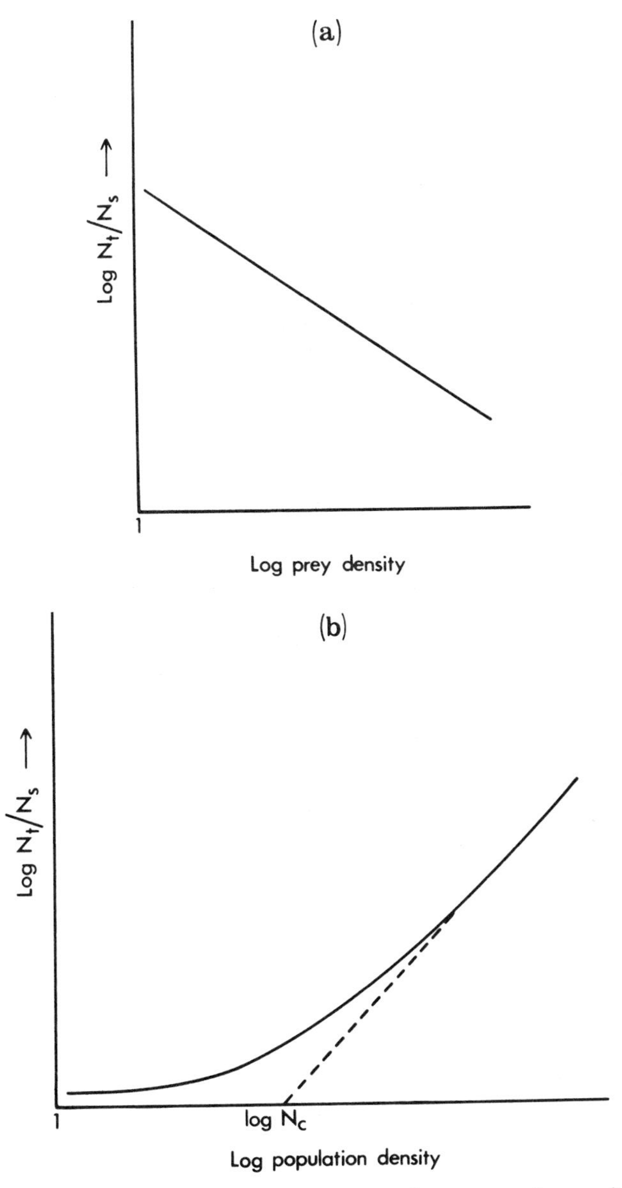

Fig. 3. Population effects expressed as k-*values (log* N_t/N_s*) in relation to density.* (a) *Functional response of a predator;* (b) *intraspecific competition effects according to Hassell's (in press) model (*N_c *= critical density).*

an aggregative response by the predator, by increasing its numbers in the area of greatest prey density (Hassell 1966; Hassell and May 1974).

2. Interference between predators or adult parasitoids (Hassell and Varley 1969; Hassell 1971*a*, *b*; Hassell and May 1973; Rogers and Hassell 1974). However, the comparable interference resulting from encounters with parasitized hosts does not increase stability (Rogers and Hassell 1974).
3. Dispersal of predators and prey and the existence of refuges for the prey (Huffaker 1958; Hassell and May 1973; Maynard Smith 1974).
4. Prey switching, where the predator switches its feeding activity to the more abundant prey (Murdoch 1969; Murdoch and Marks 1973). Steele (1974) has shown that under certain conditions prey switching may not be stabilizing.
5. Development response, where the predator is able to increase its prey consumption by virtue of its enhanced feeding capacity with increased age (Murdoch 1971).

It must be emphasized that these mechanisms may provide stability; but, at least in the models at present available, not all values of the appropriate parameters are stabilizing. Hassell and May (1973, 1974) have studied the stability properties for a number of models, involving behavioral responses, interference, aggregation, the time spent traveling between different areas (T_O), and the proportion of hosts accessible to predators. One of their simplest analyses (fig. 4*a*) shows that values of the mutual-interference constant (m) of about 0.3 will lead, unless prey reproduction (F) is especially high, to a stable equilibrium, though initially there may be oscillations of decreasing amplitude. In six laboratory studies and four field studies of different parasitoids, m has been measured and found to vary between 0.28 and 0.96 (Hassell and May 1973).

As a result of their studies, Hassell and May (1973, 1974) conclude that stability is enhanced by—

1. a small handling time relative to total searching time;
2. a measure of interference between predators (see fig. 4*a*);
3. a low prey-reproduction rate;
4. a high level of predator aggregation, its components being:
   a. The transition region (termed J by Hassell and May), when the predator rapidly increases the time spent in a unit area, coincides with the average prey density (fig. 4*b*).

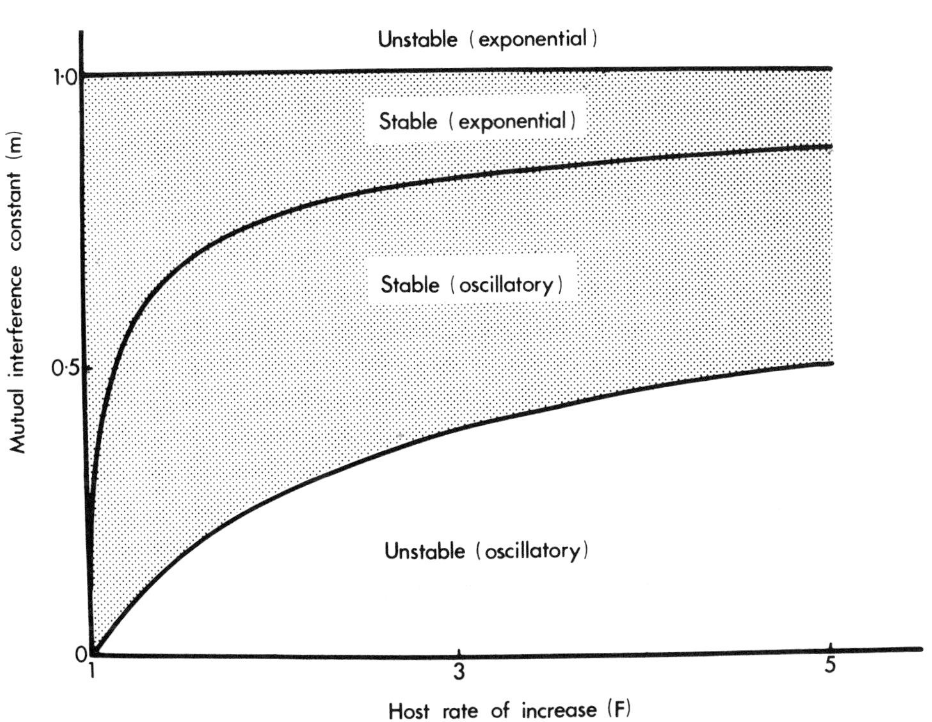

Fig. 4a. Stability boundaries for the mutual-interference constant of predators (m) *and prey reproduction (from Hassell and May 1973).*

b. The amount of time spent in areas of high prey density is greatly in excess of that spent in areas of low prey density.

c. The time spent in travel between areas (T_O) is large; hence, as dispersal is most common when prey densities are low (from *b*), this further lowers the efficiency of search at low prey densities.

d. The less clumped the prey, the greater the need (for stability) for marked predator aggregation.

These analyses confirm the general conclusion from single-species studies; namely, that, although certain interactions may have the property of moving the system towards equilibrium stability, whether they do so or not will depend on their having numerical values within a particular range. We must think of the evolution and selection of the magnitude of these parameters as paralleling the almost infinite, but equally nonrandom, variety of form and structure.

How far the system, the population, will actually move

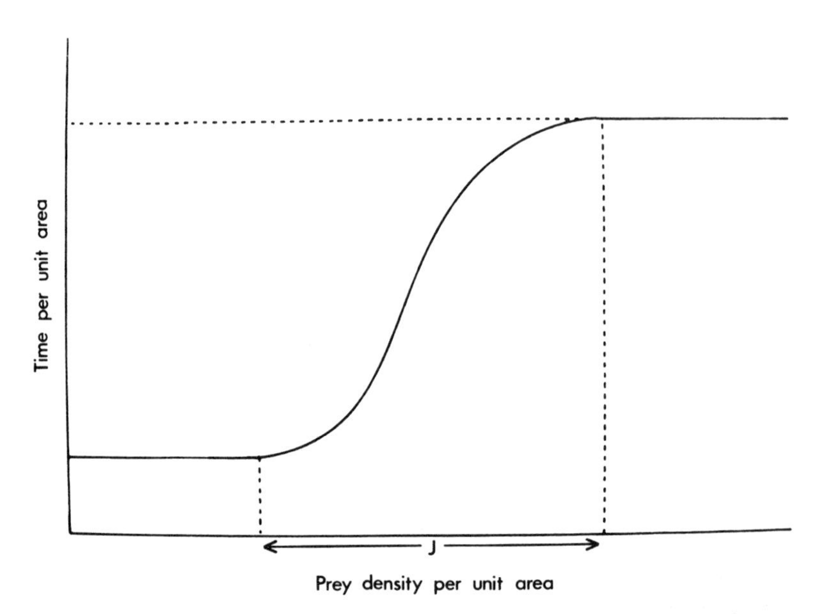

Fig. 4b. The relationship between the time spent in a unit area and prey density, showing the transition region J *(from Hassell and May 1974).*

towards stability will depend on the extent of random changes and density-independent mortalities in the environment. Thus it is now necessary to consider actual populations.

ANALYSIS OF NATURAL POPULATIONS

In any field of scientific endeavor progress demands the testing of theories against data from the real world. Experimental populations have contributed a great deal to our understanding of limited interactions; in particular, one may think of Crombie's (1945, 1946), Park's (1948, 1954), and Birch's (1948) now classic studies on stored-products beetles or of Nicholson's (1954) now almost overanalyzed observations on blowflies! More complex predator-prey studies have, similarly, been handled experimentally (e.g., Burnett 1967; Huffaker 1958; Huffaker et al. 1968; Utida 1950, 1955). Such experimental studies, whether in the laboratory or the field, will always provide especially elegant methods of investigating particular interactions; for example, the bioclimatic studies of Messenger (1964) and those of Way (1968) on intraspecific effects in aphids.

Many other ecologists have studied populations in their natural habitats where the full gamut of dynamical complexity might be hoped to reveal itself. The view may be taken that the complexity is indeed so great that an attempt to unravel it is premature; others (e.g., Way 1973) have warned against a mere Baconian gathering of data. The main problem with quantitative field studies is, to use cybernetic terms, that the level of "noise" is high. The noise arises from two sources: (a) real variation, a product of climatic variability, the stochastic nature of natural events, and the spacial heterogeneity of most environments; and (b) sampling errors. There have also been technical difficulties in calculating the number of individuals in a stage (see Richards and Waloff 1954, 1961; Southwood 1966) and in analyzing the data. Progress is continually being made, improving our techniques for sampling field populations and preparing life budgets of the type pioneered by Richards (1940, 1961). It is my belief that good field studies are essential complements to theoretical analysis, for without them one would lose both the biological insights, which can provide a most profitable starting point for theoretical analysis, and the ability to test or even substantiate one's findings in the real world (see Lawton 1974).

Analysis and Interpretation of Life Budgets

A number of different methods have been developed. Basically, there are two approaches, each of which reveals different information and makes different demands on the data.

Key-Factor Analysis. The term *key factor* was originally introduced by R.F. Morris, but the most practical method for its detection is that of Varley and Gradwell (1960), now described in several textbooks (Southwood 1966; Williamson 1972; Varley, Gradwell, and Hassell 1973). The *key factor* is that factor whose variation makes the greatest contribution to the changes in total mortality. Mortalities are expressed as k-values, the ratios of successive populations. A series of k's (k_0, k_1, k_2, . . . , k_n) is calculated for a population, and its total is the generation "mortality" (including variations in natality) K.[1] Originally, Varley and Gradwell

[1] K should not be confused with K, the carrying capacity; as the latter is undoubtedly the longer established and more widely used term, a new notation should perhaps be adopted for total generation mortality.

(1960) detected the key mortality by visual correlation with K, but Podoler and Rogers (in press) have shown that this is more precisely done by calculating the regression coefficient of each k on K; the largest regression coefficient is associated with the key factor.

As the key factor accounts for most of the change from generation to generation, this would be thought to be a disturbing factor, rather than a stabilizing, regulating one. Indeed, this is normally true, but a stabilizing factor with density-dependent action may also be the key factor if it is overcompensatory (as shown in the stability analysis above) or because of a lack of precision (a high variance) in its action (Southwood 1967) or, conversely, because it represents a very precise density-dependent factor (with b = 1) that "corrects" a number of disturbing factors. In this case the population of the next stage would remain almost constant. Satisfactory key-factor analysis demands a series of fairly complete life budgets.

One weakness of key-factor analysis arises from the role of errors in the estimates, and further problems arise when there is no stable equilibrium (Maelzer 1970; St. Amant 1970; Ito 1972). The error problem is particularly serious with insects of unstable habitats, including arable crops, where there is considerable dispersal at certain stages and thus, even if the crop is grown in the same field each year, the degree of relationship between successive generations is unknown.

The convention (see Podoler and Rogers's [in press] analysis for *Oscinella frit*) is to multiply the last generation of a season by a factor, the maximum observed increase, giving an estimate of the maximum potential natality (see table 18 of Southwood 1966). The difference between the logarithms of this estimate and of the first "real" population estimate (egg numbers) gives a value, normally designated k_0, that covers variations in adult survival, migration, and fecundity. One then determines whether this k or another contributes most to the variation in the size of K (the total generation "mortality").

It is clear that if k_0 contains a random error, arising from the assumptions about the relationship of successive populations, and if this error is sufficiently large then it will dominate the variation in K and be, perhaps spuriously, designated the key factor. Indeed, the recipe for key-factor analysis by Varley and Gradwell's method should include the general warning that confidence cannot be placed in the identification of a key factor if there is reason to believe that significantly more errors are included in the estimates contributing to this term than to other values in the life budget.

For insects living in crops this generally has the implication that the identification of key factors other than k_0 can be accepted, but not k_0 itself. This is unfortunate, as there are theoretical reasons for supposing that k_0 will often in fact be the key factor in unstable habitats. Kuno (1971) and Ito (1972) have shown how the value of b may be biased by sampling errors. Ito has cautioned against the uncritical application of key-factor analysis to multivoltine species, arguing that the key factor for, say, the first generation of a season may be different from that in the second; this point may easily be checked.

The key factors hitherto identified in insect life-budget studies are set out in table 1. A much more detailed analysis of the majority of these cases, together with work on vertebrates, is given by Podoler and Rogers (in press), whose paper will also allow the very appropriate investigation of the comparative nature and distribution of regulating factors. The examination of the key factors, as identified in table 1, suggests that—

1. For most woodland natural-habitat and stored-product herbivores, larval mortality (often parasitism or predation) is the key factor. Some sap-sucking species (*Leptopterna*, *Aleurotrachelus*) are much influenced by food quality acting on fecundity.
2. For most plant feeders of agroecosystems, variations in the numbers of adults, especially through emigration and immigration, and also possibly variations in fecundity seem to be the key factors (but, as indicated above, full reliance cannot be placed on this conclusion because of error effects).
3. For the only detritus feeder, in a temporary habitat, what appeared to be essentially scramble competition between early larvae was the factor that caused the greatest variation in population size between egg and pupa.
4. For parasites the key factor is host infection. This is clearly the dominant and critical factor in the population dynamics of many parasites.

A new approach to the analysis of life budgets is provided by Birley (unpublished), who has applied the techniques of systems analysis appropriate for time-series data. Essentially an autoregressive, moving-average method has been used that enables some recognition of sampling errors and

TABLE 1

Key factors in studies on insect populations

Habitat	Insect	Key Factor(s)[1]	Reference
		Herbivores	
FIELD CROPS, AGROECOSYSTEMS	HEMIPTERA		
Rice, Japan	*Nephotettix*	+Adult mortality, migration, and fecundity variations	Kiritani et al. 1970 (this paper)
Rice and vegetables, Japan	*Nezara*	+Adult mortality, migration, and fecundity variations	Kiritani 1964 (this paper)
Sugar, Jamaica	*Saccharosydne*	Adult mortality and migration	Metcalfe 1972
	COLEOPTERA		
Potato, etc., Japan	*Epilachna*	+Mortality of 4th larval → adult stages	Iwao 1971
Potato, U.S.A.	*Leptinotarsa*	+Migration	Harcourt 1971
	DIPTERA		
Oats, U.K.	*Oscinella*	+Adult mortality, migration, and fecundity variations	Southwood and Jepson 1962 (Southwood 1967; Podoler and Rogers, in press)
Cabbage, Canada	*Erioischia*	Adult mortality, migration, and fecundity variations	Mukerji 1971 (Benson 1973*a*)
Cabbage, U.K.	*Erioischia*	+Egg predation	Hughes and Mitchell 1960 (Benson 1973*a*)
WOODLAND	LEPIDOPTERA		
Oak, U.K.	*Operophtera*	"Winter disappearance" (adult, egg, and early larval survival and fecundity)	Varley and Gradwell 1968
Oak, Canada	**Operophtera*	Mortality of larvae (parasitism)	Embree 1965
Oak, U.S.A.	*Porthetria*	Larval and pupal mortality	Bess 1961 (this paper)
Oak, Japan	*Hyphantria*	+Mortality of mature larvae (including parasitism)	Ito and Miyashita 1968 (this paper)
Pine, Netherlands	*Bupalus*	Mortality of larvae (parasitism, predation, and disease)	Klomp, 1966*a*, *b*, 1968
Larch, Switzerland	*Zeiraphera*	Fecundity, egg, pupal, and adult survival	Auer 1968 (Varley and Gradwell 1970)
Spruce, Canada	*Choristoneura*	+Mortality of later larvae	Morris 1963
Spruce, Canada	*Acleris*	Mortality of larvae (parasitism)	Morris 1959 (Varley, Gradwell, and Hassell 1973)
Conifer, Japan	*Dendrolimus*	Mortality of larvae	Kokubo 1965 (this paper)
	COLEOPTERA		
Fir, U.S.A. (NW)	*Scolytus*	+Mortality from egg to adult	Berryman 1973

Habitat	Insect	Key Factor(s)[1]	Reference
	HEMIPTERA		
Olive, California	*Parlatoria*	Parasitism of larvae by *Aphytis*	Huffaker and Kennett 1966
Eucalyptus, Australia	*Cardiaspina* (endemic)	Variations in adult survival and fecundity	Clark 1964 (this paper)
	PSCOPTERA		
Algae on larch, U.K.	*Mesopsocus* (2 spp.)	+Predation of larvae	Broadhead and Wapshere 1966
SCRUBLANDS AND GRASSLANDS	COLEOPTERA		
Broom, U.K.	*Phytodecta*	Larval mortality	Richards and Waloff 1961
	LEPIDOPTERA		
Rumex, U.K.	**Lycaena*	Mortality of egg and first instar	Duffey 1968
Senico, U.K.	*Tyria*	Larval mortality, especially starvation	Dempster 1971 (and pers. comm.)
	HEMIPTERA		
Holcus, U.K.	*Leptopterna*	(1) Variations in fecundity, (2) mortality, predation, and accident, late larvae	McNeill 1973
Viburnum, U.K.	**Aleurotrachelus*	Variations in fecundity	Southwood and Reader (pers. comm.)
STORED PLANT PRODUCTS	LEPIDOPTERA		
Flour	*Anagasta*	Parasitism of larvae	Hassell and Huffaker 1969
Flour	*Plodia*	Parasitism of larvae	Podoler 1974
Wheat tippings	*Ephestia*	Variations in fecundity, egg, and early larval mortality	Benson 1974
		Detritus Feeders	
	DIPTERA		
Artificial water containers, Thailand	*Aedes*	+Mortality of early larval stages	Southwood et al. 1972
		Parasites	
	DIPTERA		
Oak (host: Lep. larvae)	*Cyzenis*	Egg mortality (failure to be ingested by host)	Hassell 1968
	HYMENOPTERA		
Stored products (host: Lep. larvae)	*Bracon*	Variations in fecundity	Benson 1973*b*

NOTE: * means introduced; + means some assumptions in analysis.

[1]Where several factors are given, these, unless numbered, are conglomerates.

their effects, as well as direct and delayed density dependence and the delayed effect of disturbances.

Predictive Analysis. The predictive factor is the one that is most closely correlated with the ensuing population. This was the initial concept of key factor as used by Morris, but, as Podoler and Rogers (in press) point out, it is not necessarily the same as the factor accounting for change in "mortality." A method to assess the relative importance of natality and mortality in determining subsequent population size was described by Southwood (1967). The changes in the population are based on the equation:

$$\log P_R = \log P_E - M$$

where P_E = egg population and P_R = resulting (adult, pupal, or late larval) population. The method is particularly appropriate for highly mobile species (pests of crops) where, as discussed above, a series of meaningful life budgets is difficult to obtain.

The results of predictive analyses, using this method, of a large number of the available sets of data from field populations of insects are given in table 2. High values of the coefficient of determination (r^2 is the square of the correlation coefficient) imply that the given proportion of the variation in the resulting population is determined by the size of the egg population, itself normally a product of two pathways of population change: migration and natality. By analogy, very low values of r^2 may be taken as indicating the important role of mortality between the egg and the "resulting population" (normally adult, but sometimes late larvae or pupae).

It will be seen that in many of the arable crops natality provides an excellent prediction of future population size; in other words, variations in mortality from generation to generation are not significant; nor, as natality varies (when P_E covers a range), does mortality regulate (act as a major density-dependent factor). The same applies to the introduced insects on nonnative plants and to the red locust, *Nomadacris*.

In other habitats natality is not a good predictor; that is, mortality is a more significant pathway and may include density-dependent action. Little reliance can be placed on the exact magnitude of the nonsignificant values of r^2. Certain variations are, however, noteworthy: mortality is more important in vegetable crops than in cereal crops. In the corn borer (*Ostrinia*) it becomes more important in the second generation, whereas in the moth *Dendrolimus* the role of natality appears to increase in the endemic situation. The analysis for *Cardiaspina* confirms Clark's conclusion that in the endemic situation adult mortality is critical and subsequent mortalities less variable.

Specific Models of Natural Populations

Models are of two types, general and specific (Conway and Murdie 1972). The general models explore the dynamic properties of the system, and models of this type have been discussed in the section on theoretical stability analysis. It has been shown that they reveal features about the behavior of the systems, features that were by no means intuitively obvious.

Specific models are concerned with particular populations, and they may serve three functions: instructive, predictive, and analytical.

Instructive. The actual process of constructing the model can highlight gaps in the research program. The development of the flow chart and other initial stages of model construction should ideally be an integral part of research planning.

Predictive. The models are normally designed to predict the effects of changes in the variables. This is especially valuable for pest populations, where these models are frequently used (Watt 1963*a* and *b*; Conway and Murdie 1972; Conway 1973). Such models often work "back" from the observed events towards the identification of the form and magnitude of the major parameters. Then these may be varied so that the effect of various control strategies can be assessed.

As Conway et al. (in press) have stressed in their study of the sugar cane froghopper, it is essential that economic considerations be introduced if the model is to provide insights of practical relevance. On this criterion—that is, the improvement of pest-management techniques—models have not so far been very successful (Conway 1973), but progress is likely with the better definition of objective functions, which will incorporate economic considerations and the use of the appropriate optimization techniques such as dynamic programming (Shoemaker 1973*a*, *b*, *c*; Conway et al., in press).

Existing models have been more concerned with the prediction of population change within a year, with answering the question "under what climatic (and other) conditions is a pest outbreak likely?" The first extensive work of this type was for the spruce budworm (Morris 1963), and Campbell (1967)

TABLE 2

Predictive analyses of population data for various insects

Habitat	Insect	r^2%[a]	Source
ARABLE CROPS	HERBIVORES		
Oats	*Oscinella frit*	97[xxx]	Southwood and Jepson 1962
Maize	*Ostrinia nubilalis* (1st gen.)	96[xxx]	Chiang and Hodson 1959
Maize	*O. nubilalis* (2nd gen.)	49[x]	Chiang and Hodson 1959
Rice	*Nephotettix ceneticeps*	66[x]	Kiritani et al. 1970
Potato	*Leptinotarsa decemlineata*	99[xxx]	Harcourt 1971
Vegetables, rice	*Nezara viridula*	46	Kiritani 1964
Cabbage, Canada	*Erioischia brassicae*	42	Mukerji 1971
Cabbage, U.K.	*E. brassicae*	21	Hughes and Mitchell 1960
Cabbage, Holland	*E. brassicae*	11	Abu Yamen 1960
Cabbage, U.S.A.	*Trichoplusia ni*	3	Elsey and Rabb 1970
Sugar	*Diatraca*	16[x]	Pickles 1936
INTRODUCED INSECTS ON NONNATIVE PLANTS			
Olive, U.S.A.	*Parlatoria oleae* (before biological control)	95[xxx]	Huffaker, Kennett, and Finney 1962
Viburnum, U.K.	*Aleurotrachelus jelinekii*	92[xxx]	Southwood and Reader (unpublished)
ARID LANDS			
Various	*Nomadacris septemfasciata*	79[xx]	Stortenbecker 1967
MARSHLAND AND GRASSLAND			
Rumex	*Lycaena dispar*	20	Duffey 1968
Grass	*Chorthippus parallelus*	26	Richards and Waloff 1954
Grass	*C. brunneus*	1	Richards and Waloff 1954

Grass	*Leptopterna dolabrata*	35	McNeill 1973
SCRUBLAND			
Sarothamnus	*Arytaina sarothamni*	6	Watmough 1968
Sarothamnus	*Phytodeta olivacea*	0.1	Richards and Waloff 1961
Senico	*Tyria jacobaeae*	2	Dempster, in press
WOODLANDS			
Oak, U.S.A.	*Porthetria dispar*	23	Bess 1961
Plane, Japan	*Hyphantria cunea*	45	Ito and Miyashita 1968
Pine, Japan	*Dendrolimus spectabilis* (endemic)	8	Kokubo 1965
Pine, Japan	*D. spectabilis* (epidemic)	40^{x}	Kokubo 1965
Pine, Netherlands	*Bupalus piniarius*	4	Klomp 1966*b*
Fir, U.S.A.	*Scolytus ventralis*	0.1	Berryman 1973
Eucalyptus, Australia	*Cardiaspina albitextura*		Clark 1964
	(endemic 1st gen.)	60^{xx}	
	(endemic 2nd gen.)	59^{xx}	
	(endemic 3rd gen.)	92^{xxx}	
	(epidemic all gen.)	74^{x}	
	PARASITE OF DEFOLIATORS		
Oak, U.K.	*Cyzenis albicans*	29	Hassell 1968
	ALGAL EPIPHYTE FEEDERS		
Larch, U.K.	*Mesopsocus immunis*	64^{x}	Broadhead and Wapshere 1966
Larch, U.K.	*M. unipunctatus*	50	
ARTIFICIAL WATER CONTAINERS	DETRITUS FEEDERS		
Detritus, etc. Thailand	*Aedes aegypti*	0.3	Southwood et al. 1972
STORED PRODUCTS	HERBIVORE		
Wheat toppings	*Ephestia cautella*	11	Benson 1974

NOTE: r is significant at the ^{x}0.1 level, ^{xx}0.5 level, and ^{xxx}0.001 level.

[a] For resulting population with egg numbers.

has developed a model for the gypsy moth. An apparently successful model for the cabbage aphid and its parasites in Australia permitted the correct prediction not only of the level of pest population in relation to early-season parasitism, but also of the mean fecundity of the parasite in another part of its range (Europe)(Hughes and Gilbert 1968; Gilbert and Hughes 1971).

Analytical. One cannot be completely confident that a model that provides a reasonable fit is actually a realistic representation of the interaction of all the parameters and not merely an effective mimic; obviously, the greater the range of independent data against which it is tested, the greater confidence that can be placed in it. A detailed model can provide considerable insight into the role of different components in the dynamics of the population in a given season. Gilbert and Gutierrez (1973) make a powerful plea for the reasonably complete analysis by modeling of numerous sets of field data, for they believe that by such means "generalisations (should they exist)" will be revealed.

These analytical models do, of course, grade into the type of models that arise from key-factor analysis; indeed, the spruce-budworm work is the starting point of both approaches. However, the key-factor analyses described above are based largely on the comparison of successive population figures (i.e., life budgets) and the determination of year-to-year changes. The analytical aphid models are based on a combination of field and laboratory knowledge, together with climatic and phenological inputs, and essentially reveal the development of the population within a season; that is, the causal mechanisms underlying the predictions (Hughes and Gilbert 1968; Gilbert and Hughes 1971; Gilbert and Gutierrez 1973; Gutierrez et al. 1974).

Undoubtedly, such specific simulation models will, in due course, reveal further insights into basic population dynamics (particularly the evolutionary pressures), but, by comparison with key-factor and predictive analyses discussed above, so few have hitherto been undertaken that even a preliminary synthesis would be premature.

THE THREE-DIMENSIONAL FRAMEWORK OF POPULATION DYNAMICS

The Population-Growth and Population-Density Axes

The simplest representation of population growth in a limited environment is provided by the classic logistic curve in which, after an effectively exponential increase, the growth rate is successively dampened with increasing density and, eventually, asymptotically approaches the equilibrium level. Few real populations, particularly those of insects, seem to behave in this way, and as pointed out in the first section these observations caused many ecologists to reject the concept of stability and doubt the significance of density dependence.

Others, notably Milne (1957), have recognized that there appeared to be levels from which the population rose from time to time, to outbreaks. Clark (1964) crystalized these ideas, in relation to his study of the psyllid *Cardiaspina*, referring to endemic and epidemic levels. A series of the theoretical ideas will now be compiled, and I hope to show how, combined, they produce a general model that conforms with the experience of the field ecologist and yet embodies both stability and change.

The models described in the section on theoretical stability analysis are all deterministic, but in the real world random environmental variation occurs. Diamond (in press) has considered the effect of introducing random variation in two parameters, host-reproduction rate and parasite-interference constant; he found that the conditions for stochastic stability were not very different from those predicted by the deterministic model. One might summarize that the stochastic nature of the real world generally results in no more than a measure of "fuzziness" around the stability points.

In nature, populations also suffer from perturbations, major disturbances. Holling (1973) has focused attention on the concept of resilience in ecological systems. *Resilience* is the extent to which the system can recover from a change in its state variables arising from perturbations. Some population systems return, quickly or slowly, after perturbation to the same equilibrium value; these are resilient, compared with other systems that after perturbation move to a new set of dynamic properties and a new equilibrium. When a natural grassland is plowed and cultivated (a perturbation), some insects (e.g., soil dwellers) eventually regain their former population levels; others that are less resilient to such perturbation either become extinct or establish new, and often much higher, equilibrium levels. One can visualize domains of stability, areas in phase space, that represent the resilience of a population system (or ecosystem), but their quantification is not easily conceived (May 1974).

These theoretical concepts and those explored in the

section on theoretical stability analysis suggest that—

1. a population showing stability and acting under density-dependent regulation may not be constant; it may oscillate with stable limit cycles or show less regular fluctuations;
2. systems may have domains of stability and, when they are perturbed beyond one domain, they will pass to another; i.e., the population establishes itself at a new level.

The two main types of regulatory mechanisms are natural enemies and intraspecific competition. Although the studies of Hassell, May, Murdoch, and others (cited above) have shown categorically how natural-enemy action may, through certain features, stabilize prey populations (and hence Milne [1957] erred in regarding them as inherently imperfectly density dependent), it is intuitively obvious that certain rapid rates of prey density (perhaps by immigration) could saturate natural-enemy control. Indeed, Holling (1966) showed that the functional response typical of invertebrate predators would be of an inversely density-dependent (destabilizing) type (fig. 3*a*). However, switching, aggregation (numerical response), and possibly the learning of a prey type (Samson-Boshuizen, Lenteren, and Bakker 1974) may lead to a humped response curve. A total response of this type has been utilized by Fransz (1974) in his study of an acarine predator system and is represented graphically in figure 5.

Competition models have often been linear in form, yet biologically it seems sound to consider that there is a critical population level (N_c) at which scramble competition starts to operate. Below this level there may be some measure of contest competition; indeed, as Hassell (in press) points out, these two may be regarded as the extreme ends of a range. He has developed a new model for intraspecific competition in discrete generations of the form:

$$N_{t+1} = \lambda(1 + aN_t)^{-b} N_t \qquad (3)$$

where λ is the net rate of increase, b is the density-dependent moderator, and $a = 1/N_c$.

This model will describe a curve of the type shown in figure 3*b* for the relationship of mortality (expressed as a k-value, i.e., log N_t/N_s) to density. Its apparent realism, by introducing a critical density, is confirmed by the wide range of natural data to which it can be fitted (Hassell, in press).

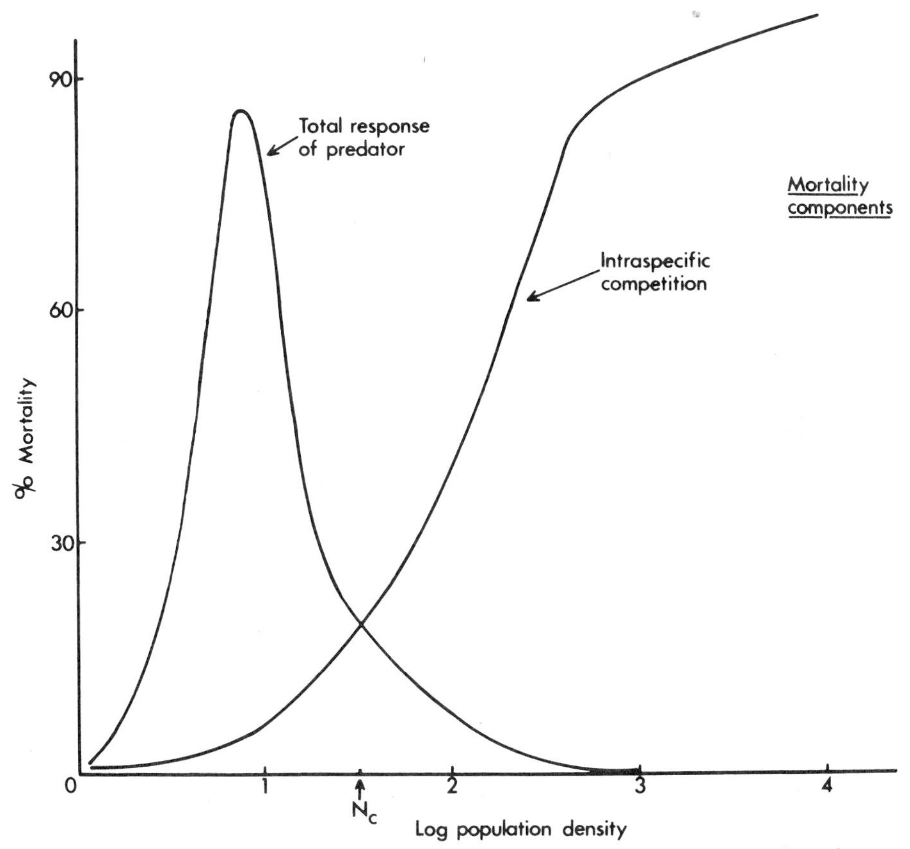

Fig. 5. Percentage-mortality curves in relation to density for the total response of a predator (after Fransz 1974) to a prey and for intraspecific competition (in the prey) based on the model of Hassell (in press)(eq. 3 in text) where N_c = *330 and* b = *2.*

These two regulatory factors may be combined graphically to produce a composite mortality curve (fig. 6). The total response of a predator, of the form described by Fransz (1974), operates over low densities (peak percent of predation at nine per unit area); at higher densities Hassell's competition model becomes dominant (N_c = 330, b = 2)(fig. 5). To complete the parameters of this hypothetical population (which is, however, based on field-tested relationships), a natality curve has to be inserted. The natality-density

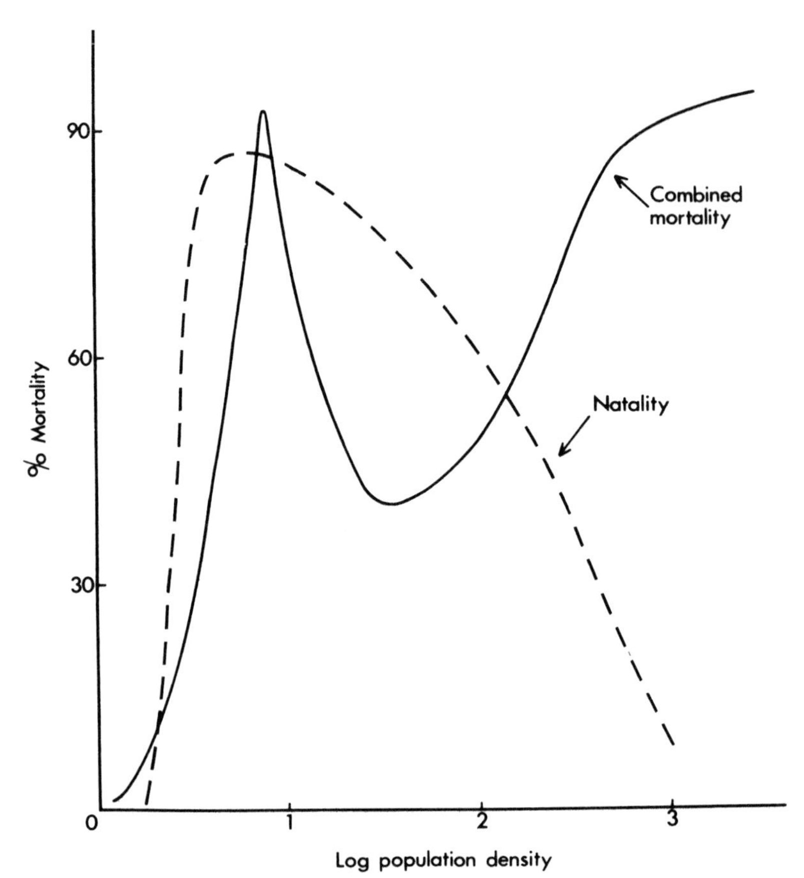

Fig. 6. The total mortality-density relationship from a combination of predation and intraspecific effects, as shown in figure 5, together with a natality-density curve (based on the form found in cabbage aphid by Way [1968]).

curve is based on that for the cabbage aphid (*Brevicoryne brassicae*), as described by Way (1968). Mortality and natality values may be read off and a composite graphical model of the basic "reproduction curve" type used by Takahashi (1964) and Holling (1973) constructed (fig. 7).

Several important points may be recognized in figure 7:

1. The extinction point *X*. Once the population is perturbed to this level or below it, extinction is inevitable, due to the presence of an "Allee effect" in the natality curve; i.e., natality falls with

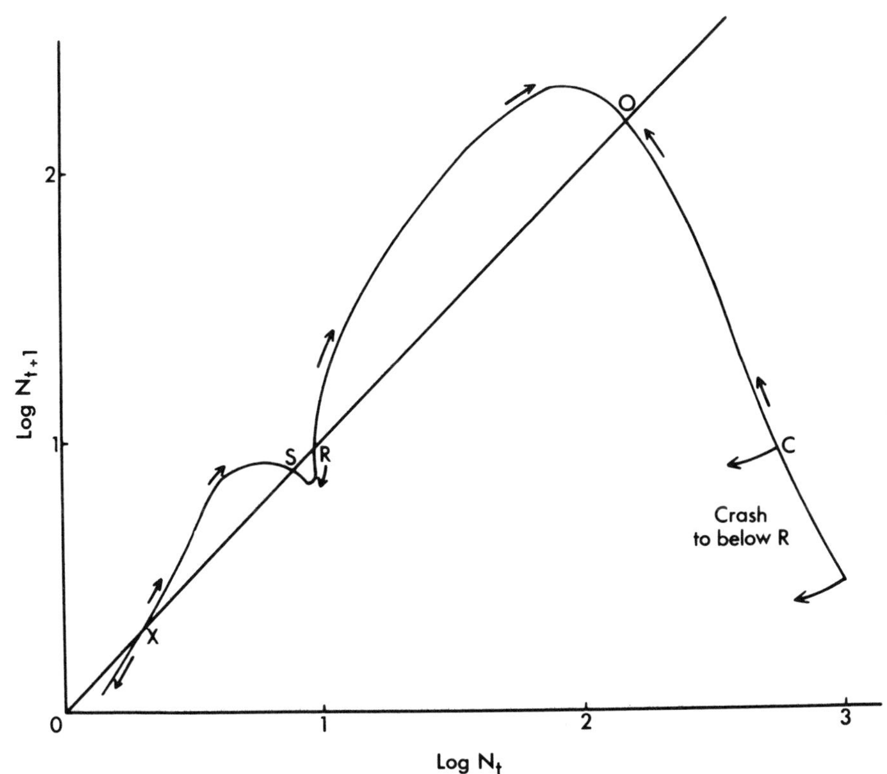

Fig. 7. A general representation of the population-growth and population-density relationship based on the natality and mortality curves shown in figure 6 (X = extinction point, S = stable equilibrium point, R = release point, O = oscillations equilibrium point around which populations move in stable limit cycles, C = crash point beyond which population returns to lower equilibrium level).

decreasing density below a certain optimum level.

2. A stable equilibrium point *S*, determined by the interaction of the predator and natality. Population below it will rise to it; those just above it will fall back to it.
3. A release point *R*, at which the predator can no longer contain population growth. Once beyond this point, population continues to rise, intraspecific mortality and natality effects increasingly operating to dampen

growth. This release point corresponds with Voûte's (1946) concept of "escape."

4. An oscillations equilibrium point *O*, around which the population will move in stable limit cycles. This arises because *b* in the Hassell model was given the value of 2; hence this mortality is overcompensating and a two-point stable limit cycle will be produced.
5. The crash point *C*, beyond which, if the population is perturbed (by excessively favorable conditions for reproduction, survival, or immigration), the population crashes to a level below which predator control again operates.

The region *X-S-R* may be considered as the endemic population level and that of *R-O-C*, as the epidemic level of Clark (1964). The small range of population densities in the region *S-R* arises solely from the form of the underlying natality and mortality curves. One might consider that it gives the equilibrium system of *S* a small domain of stability. Whether an equilibrium point is *S* or *O* will depend entirely on the value of *b* in its region (see section on theoretical stability analysis). Representative population traces are shown in figure 8, from which it will be seen that very different patterns of fluctuation arise, depending on the extent of "noise" (random variation in the birth and death rates).

This model appears to represent adequately the pattern typical of many field populations. Clark's (1964) studies on the eucalyptus psyllid *Cardiaspina albitextura* (see also Clark et al. 1967) showed that during the endemic phase numbers were regulated principally through density-dependent predation of the adults by a range of general predators, mostly birds. Mortality of the immature stages was generally independent of density (see table 2): however, when these mortality factors (especially parasites) were depressed by weather conditions, the ability of bird predation to respond to the rise in density was soon exceeded (i.e., the population passed through a small *S-R* zone). The mean number of adults per shoot rose from around two to nearly fifty, the epidemic level. Extreme scramble competition for food occurs as the foliage becomes damaged; larvae perish from starvation and female oviposition may be limited by the small amount of suitable foliage. In terms of the model, the point *O* is close to the point *C* to which the population usually rises, then returning to the low endemic level.

A similar example, also from Australia, is provided by the eucalyptus phasmatid, *Didymuria violescens* (Readshaw 1965).

Endemic population densities of about 10-20 per 100 ft^2 are regulated by natural enemies (again principally birds); climatic perturbations are followed by rises in populations to 45-55 per 100 ft^2, which seems to constitute the release-point density; thereafter densities of up to 125 per 100 ft^2 are attained when intraspecific competition (through damage to the host plant) becomes severe.

The history of the European spruce sawfly (*Diprion hercyriae*) in Canada effectively represents movement down through the model. Achieving epidemic proportions after its introduction from Europe, its numbers were eventually reduced, primarily by a virus disease, to very low endemic levels where introduced parasites now seem to be effective regulating factors (Neilson and Morris 1964). Successful biological control can generally be regarded as moving the population from the epidemic to the endemic phase.

In cereal fields in Britain it seems that, if there are significant numbers of predators present early in the spring, subsequent aphid outbreaks are less likely (Potts 1970; Potts and Vickerman 1974); presumably the aphid populations are kept in the zone *X-R*. Densities during outbreaks are some hundredfold greater.

The perturbations that cause populations to pass beyond the release point are frequently climatic; Ito (1961) provides a number of examples. Another potent cause is the use of insecticides: natural enemies are adversely affected and new population levels are achieved; new pests appear (Massee 1952; Ripper 1956; Entwistle, Johnson, and Dunn 1959; Smith 1970; Conway 1971).

The situation represented by the region *R-C* in the model is characteristic of many species of unstable habitats, e.g., locusts and blowflies. Density dependence is probably overcompensatory; the highest value of b recorded from the field is for scramble competition in *Aedes aegypti* larvae in artificial water containers (Southwood et al. 1972). One of the most important pathways of population change at levels close to or above the carrying capacity is migration; Kennedy (this volume) has reviewed our knowledge of this ecologically significant behavior.

The remaining situation shown by the model is the one in which the density of the stability point *O* is not sufficiently above the carrying capacity of the environment to cause its permanent destruction, but the population oscillates around that level. The larch moth, *Zeiraphera diniana*, in the Engadine, Switzerland (Baltensweiler 1968, 1971), and the tent caterpillar, *Hyphantria cunea*, in Canada and Japan

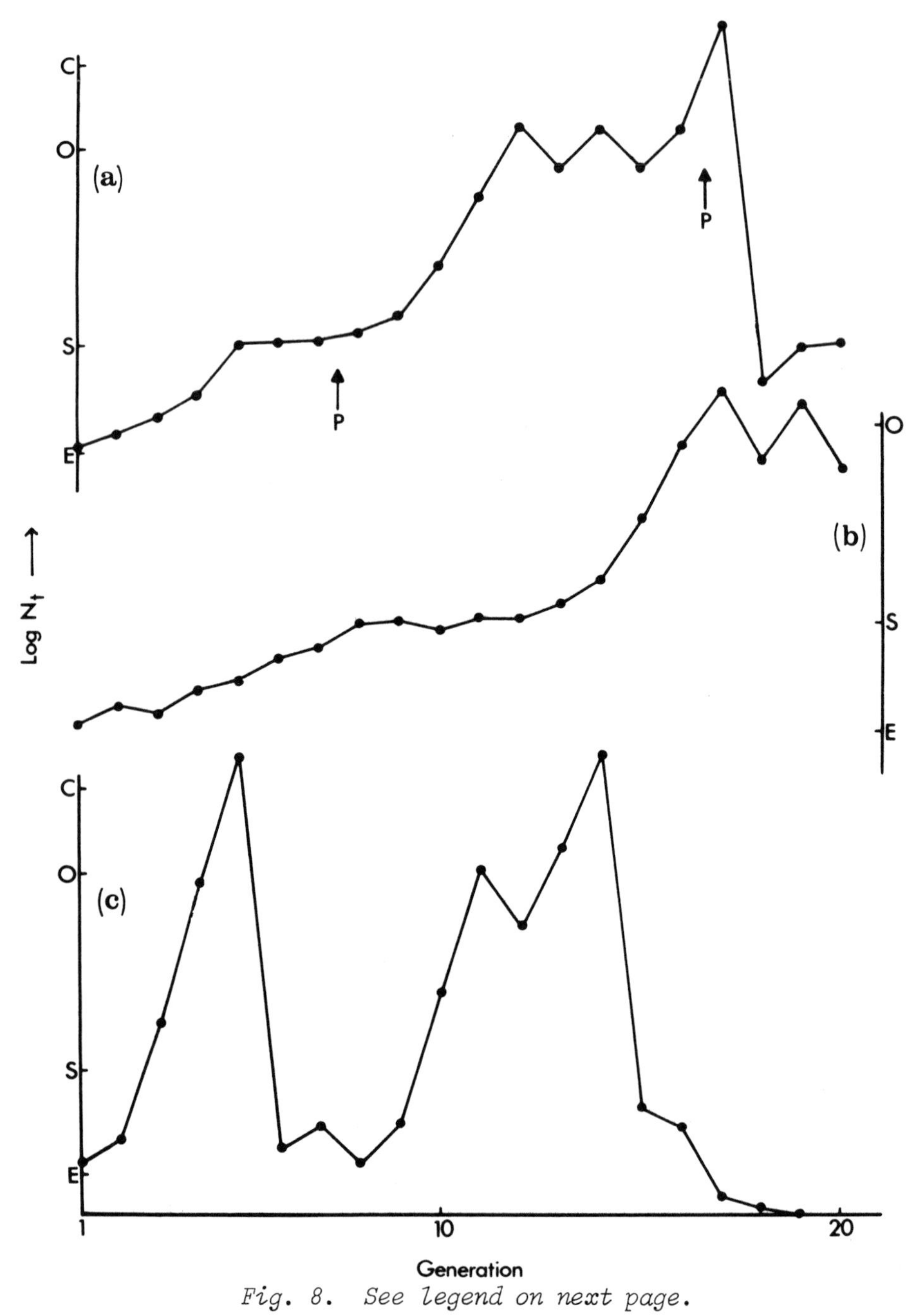

Fig. 8. See legend on next page.

(Morris 1964; Ito, Shibazaki, and Iwahashi 1969) appear to provide excellent examples of oscillations at high density.

The data are not available to determine with certainty if any natural populations could exhibit chaotic behavior solely through the action of density-dependent regulation (May 1975) (see the section on theoretical stability analysis and fig. 2). A putative example is provided by the social wasps, especially *Paravespula*. Records of the number of nests (which represent one reproducing female) show apparently chaotic fluctuations (Southwood 1967); the levels of fecundity are now known to be very high (a mean queen production of over 4,000 for *P. germanica* [Spradbery 1971]); and young queens compete fiercely and fatally for nesting sites. As Spradbery (1973) points out, where both perish, severe overcompensation (a high value of b) will result.

Thus this model for population growth emphasizes that the action of stabilizing mechanisms can account for outbreaks and extinctions, fluctuations and oscillations; that is, population change in all its forms. On this "bedrock" further environmental perturbations are imposed and produce the rich variety of changes we observe (fig. 8).

The Habitat Axis

Habitats may be classified according to their stability (i.e., the degree of their permanence and predictability). The spectrum ranges from temporary pools, carrion, dung, and ruderal plants to large rivers and permanent forests (Southwood 1962*a*, *b*). It is claimed that "habitat is the template against which evolutionary pressures fashion the ecological strategy of a species; the instability-stability habitat spectrum gives rise to the r-K selection continuum" (Southwood et al. 1974). Habitat stability may be defined for any animal as τ/H (τ = generation time, and H = the length of time the habitat remains suitable for food harvesting). The habitat itself is the area accessible to the trivial movements of the food-harvesting stages. Clearly, the larger the scale, either in space or time, in which habitat is viewed, the greater its predictability; thus

Fig. 8. Population change based on the model represented in figure 7; E, S, O, *and* C *as in figure 7.* (a) *Deterministic simulation, with perturbations upwards at points marked* P; (b) *stochastic simulation with random variation: 1- to 1.5-fold increase or decrease of deterministic outcome;* (c) *stochastic simulation as* b, *but range 1-5.*

habitats of larger animals, which can control the direction of their movements over longer distances and which live longer, will be more stable than those of smaller species (Southwood et al. 1974).

It has been shown above how the population dynamics of a species is determined by a number of often interrelated parameters: the density-dependent moderator (b) which in discrete generations determines the return time, the net rate of increase (λ), and the equilibrium level (or levels) ($N^*_{1,\ldots,n}$—the highest of which will normally equal K, the carrying capacity). As animals evolve against their habitats, at different points of the instability-stability spectrum, different ecological strategies will develop from the selection of particular values for these parameters.

The fundamental requirement at the most stable end of the habitat spectrum, the K-strategist of MacArthur (1960), is that the population return rapidly to the equilibrium level after a perturbation. On the other hand, overshooting the carrying capacity will have an evolutionary penalty: the habitat will have been degraded for future generations. Thus in terms of the difference model (eq. 1) of population growth:

$$N_{t+1} = (\lambda N_t^{-b})N_t$$

b will approximate unity, but not exceed it. Southwood et al. (1974) show how this is likely to arise by maximizing the proportion surviving, lengthening the life span, and developing subtle mechanisms whereby fecundity, although normally low, may be rapidly increased if population density is lowered. This is seen in many large, long-lived vertebrates which have delayed maturity and exclude potential breeders through territorial behavior. The tropical rain-forest *Morpho* butterflies are large, territorial, and long-lived and probably represent initially K-selected insects (Young and Muyshondt 1972).

The r-strategist is an exploiter. Whittaker (1974) has recently suggested that, in addition to r (exploitation) selection, one must recognize "adversity selection." I would agree that the evolutionary answer to habitat change, where it is reversible (due to climate), may involve either a "sit-tight tactic" (i.e., diapause in insects) or one of "seeking pastures new" (i.e., migration)(see Southwood 1962*a*, especially fig. 1; Dingle 1974). However, once adversity is past, either in the old or a new habitat, the population needs to be an exploiter. To use the terminology of Whittaker, Levin, and Root (1973), a large proportion of the niche hypervolume is empty. The r-strategist will therefore strive to maximize λ, the net rate of increase. Southwood et al. (1974) show that the effective b, in the region of the equilibrium, is likely to be precipitously overcompensating. This will lead to overshooting, but, as r-strategists' habitats are unstable (Υ/H approaching unity), there is no evolutionary penalty for overexploitation.

Migration is a pathway of population change of particular significance to the r-strategist. It is found principally in those insects that are denizens of temporary habitats, providing the mechanisms for both the invasion of the ever-occurring new sites and for the evacuation of possibly over-exploited, doomed ones (Southwood 1962*a* and *b*; Sweet 1963; Johnson 1969; Dingle 1968, 1972, 1974). Kennedy (this volume) has, however, pointed out that the individual that is actually migrating is deferring its reproduction, and in this it is behaving like a K-strategist. Thus the worst penalties of overexploiting the habitat, which follow from the r-strategists' basic features, are mitigated by individuals temporarily adopting K-type behavior (postponing reproduction). This highlights the multifaceted nature of the r-K continuum.

The broad differences in population strategies proposed for r- and K-selected species may be tested against the life-budget-analysis data in tables 1 and 2. It will be seen that both key-factor and predictive analyses indicate the great significance of the number of eggs laid, itself a reflection of the changes in adult numbers through migration, in most pests of arable crops. (The corroboration by predictive analysis of the key-factor findings suggests that the latter are not artifacts of large error terms.) The work on *Aedes aegypti*, which colonizes artificial water containers, showed a highly overcompensating competitive interaction between young larvae; every adult, of course, migrates from the habitat.

In the natural habitats, mortalities due to predators and parasites are more in evidence. These habitats differ in their permanence, and some insects (e.g., *Dendrolimus*, *Cardiaspina*) reach epidemic levels occasionally. The sample is not perhaps random; entomologists tend to study species that have outbreaks, rather than those that maintain a steady level.

An intermediate habitat is the scrub, broom (*Sarothamnus scoparius*), which lives for about ten to fifteen years and thus for most univoltine species Υ/H is in the region 0.1-0.07. Studies of a large number of species living on broom (Waloff 1968*a* and *b*) have revealed the rich variety of population strategies one might expect for a dynamic, moderately unstable habitat. Migration, often related to population density (but also leading to rapid colonization of new seedlings), is a frequent feature, and mortality due to natural enemies fluctuates greatly. Many species tend to become epidemic in about the sixth year of a broom scrubland, within one or two more seasons the bushes themselves start to decline.

There is good evidence of considerable interspecific competition (Waloff 1968*a* and *b*); this indicates that "*K* evolutionary pressures" to suppress exploitative population growth until the bush is full sized will be balanced against the more extreme *r*-strategy of quick population growth as a method of seizing resources from competing species. Thus it is not surprising that the herbivorous insects tend to reach high densities somewhat before the bushes are full sized, but the mixture of endemic-epidemic levels seems particularly well suited to a species-habitat relation with intermediate values of Υ/H.

Another approach to the investigation of the role of habitat in the evolution of population parameters is to compare closely allied species. Landrahl and Root (1969) did this with two species of *Oncopeltus*; they found that population build-up was more rapid in a temperate than a tropical species. It must be emphasized that temperate habitats are not necessarily more unstable than tropical ones (e.g., a temperate oak woodland is more stable than a semiarid grassland of Africa); each habitat should be considered for each animal species against the Υ/H ratio criterion.

The Framework

The accumulating evidence from natural populations therefore supports the conclusion that a general representation of population dynamics may be obtained by arranging the variations of types of population growth-density relationships along a third axis, that of habitat stability. The extreme *r*- and *K*-strategies (as shown by Southwood et al. 1974) represent the front and near boundaries of the landscape (fig. 9). The "natural enemies ravine" represents stable equilibrium points and their immediate domains of stability. It runs diagonally, but is absent at both boundaries to habitat stability.

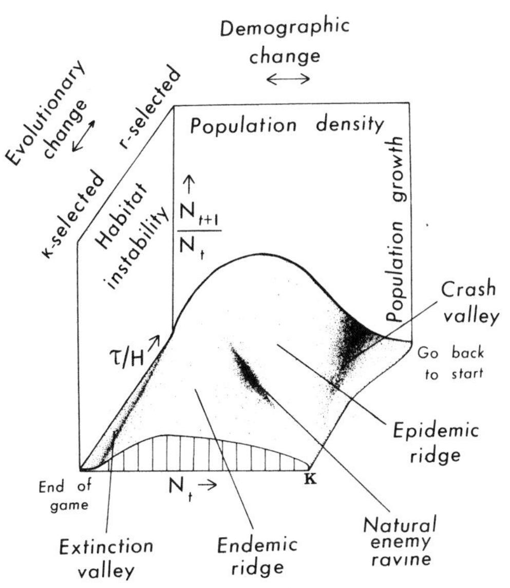

Fig. 9. A synoptic view of the landscape of population dynamics with axes: population growth (N_{t+1}/N_t)*, population density* (N_t)*, and habitat stability* (τ/H).

In unstable habitats (e.g., aphids in crops) natural enemies can stabilize growth only if they act early in population growth; otherwise their action is often inversely density-independent and involves essentially mopping up, as the now massive population moves across the "crash valley." In more stable habitats the endemic level will be relatively higher and the proportional increases, lower; compare a cereal pest, fruit fly, which increases 50-fold in one generation, with the eucalyptus stick insect, *Didymuria*, which increases 10-fold during outbreaks. The release face, on the "far side" of the ravine, will be steep if predator action is density dependent with b = 1; its "height" will depend on the distance (in terms of population density) between the stable equilibrium (which lies in the bottom of the ravine) and the release point (i.e., S and R of fig. 7). Beddington and Free (in press) have shown, by theoretical analysis, how stable coexistence in a predator-prey system is possible well below the prey carrying capacity.

For the species of unstable habitats, the "crash valley" is broad and widens beyond the limits of the carrying capacity; the population "runs off the board" here and goes back to the start; i.e., migrants colonize new habitats. Higher up, the valley becomes bounded as density-dependent moderation is less extremely overcompensating, and this may be considered to represent the region where populations (e.g., *Zeiraphera*) oscillate. As we move, on this face, towards the stable end of the habitat axis, the edge of the landscape approximates the contour 1 of the population growth axis; that is, the dynamics of a K-strategist near equilibrium (Southwood et al. 1974). The "extinction valley" becomes deeper with habitat stability; that is, X occurs at progressively higher population densities. It is "the end of the game" for populations that fall into it.

This relief map of the landscape, across which populations and species play the existential game of population dynamics, is somewhat allegorical in places and certainly over-simplified. Undoubtedly, some species have several stability points; that is, other ravines or valleys will be present. Perhaps another dimension will need to be added for the scale of the habitat—a problem considered by Boer (1968), Brussard and Ehrlich (1970), and others. However, it would seem that stochastic events would be more dominant in smaller-scale habitats, effectively moving the dynamical behavior of the population from, for example, that represented in figure 8*b* to that in figure 8*c*. In the final analysis, figure 9 may appear simplistic, as ancient maps of the world do now, but in the meantime I hope it will assist us in our thinking. Clearly, stability and change are not alternative explanations, but interrelated phenomena in the game of existence.

ACKNOWLEDGMENTS

I am deeply indebted to M.P. Hassell for critically reading the manuscript and for much assistance in its development; other colleagues have discussed particular points and helped in various ways, notably G.R. Conway, P. Diamond, J.S. Kennedy, K. Majer, R.M. May, P. Reader, and N. Waloff. I am most grateful to them and others who have generously made "in press" papers and unpublished data available to me.

REFERENCES

Abu Yamen, I.K. 1960. Natural control in cabbage root fly populations and influence of chemicals. Meded. Landbouwhogesch. Wageningen 60:1-57.

Andrewartha, H.G., and L.C. Birch. 1954. The distribution and abundance of animals. University of Chicago Press, Chicago.

Auer, C. 1968. Erste Ergebnisse einfacher stochastischer Modelluntersuchungen über die Ursachen der Populationsbewegung des grauen Lärchenwicklers *Zeiraphera diniana* Gn. (= *Z. griseana Hb.*) im Oberengadin, 1949/66. Z. Angew. Entomol. 62:202-35.

Bakker, K. 1964. Backgrounds of controversies about population theories and their terminologies. Z. Angew. Entomol. 53:187-208.

Baltensweiler, W. 1968. The cyclic population dynamics of the grey larch tortrix *Zeiraphera griseana* Hübner (= *Semasia diniana* Guenée) (Lepidoptera: Tortricidae), p. 88-97. *In* T.R.E. Southwood [ed.], Insect abundance. Roy. Entomol. Soc. London Symp. 4. Blackwell Scientific Publ., Oxford.

———. 1971. The relevance of changes in the composition of larch bud moth populations for the dynamics of its number, p. 208-19. *In* P.J. den Boer and G.R. Gradwell [eds.], Dynamics of populations. Proc. Advan. Study Inst. Dyn. Numbers Pop., Oosterbeek, Netherlands, 1970. Pudoc, Wageningen.

Beddington, J., and C.A. Free. Age structure, density dependence and limit cycles in predator-prey interactions. J. Theor. Pop. Biol., in press.

Benson, J.F. 1973*a*. Population dynamics of cabbage root fly in Canada and England. J. Appl. Ecol. 10:437-46.

———. 1973*b*. Intraspecific competition in the population dynamics of *Bracon hebetor* Say (Hymenoptera: Braconidae). J. Anim. Ecol. 42:105-24.

———. 1974. Population dynamics of *Bracon hebetor* Say (Hymenoptera: Braconidae) and *Ephestia cantella* (Walker) (Lepidoptera: Phycitidae) in a laboratory ecosystem. J. Anim. Ecol. 43:71-86.

Berryman, A.A. 1973. Population dynamics of the fir engraver, *Scolytus ventralis* (Coleoptera: Scolytidae). 1. Analysis of population behaviour and survival from 1964-1971. Can. Entomol. 105:1465-88.

Bess, H.A. 1961. Population ecology of the gypsy moth *Porthetria dispar* L. (Lepidoptera: Lymantridae). Conn. Exp. Sta. Bull. 646.

Birch, L.C. 1948. The intrinsic rate of natural increase of an insect population. J. Anim. Ecol. 17:15-26.

Birley, M.H. The assessment of density dependence from time series of animal populations. Unpublished ms.

Boer, P.J. den. 1968. Spreading of risk and stabilization of animal numbers. Acta Biotheor. 18:165-94.

Broadhead, E., and A.J. Wapshere. 1966. *Mesopsocus* populations on larch in England—the distribution and dynamics of two closely-related coexisting species of *Psocoptera* sharing the same food resource. Ecol. Monogr. 36:327-88.

Brussard, P.F., and P.R. Ehrlich. 1970. Contrasting population biology of two species of butterfly. Nature, London 227:91-92.

Burnett, T. 1967. Aspects of the interaction between a chalcid parasite and its aleurodid host. Can. J. Zool. 45: 539-78.

Campbell, R.W. 1967. The analysis of numerical change in gypsy moth populations. Forest Sci. Monogr. 15:1-33.

Chiang, H.C., and A.C. Hodson. 1959. Population fluctuations of the European corn borer *Pyrausta nubilalis* at Waseca, Minnesota, 1948 to 1957. Ann. Entomol. Soc. Amer. 52:710-24.

Clark, L.R. 1964. The population dynamics of *Cardiaspina albitextura* (Psyllidae). Aust. J. Zool. 12:362-80.

Clark, L.R., P.W. Geier, R.D. Hughes, and R.F. Morris. 1967. The ecology of insect populations in theory and practice. Methuen, London.

Conway, G.R. 1971. Pests of cocoa in Sabah and their control. Dep. Agr., Malaysia, Kuala Lumpur.

———. 1973. Experience in insect pest modelling: a review of models, uses and future directions, p. 103-30. *In* P.W. Geier, L.R. Clark, D.J. Anderson, and H.A. Nix [eds.], Insects: studies in population management. Ecol. Soc. Aust. Mem. 1, Canberra.

Conway, G.R., and G. Murdie. 1972. Population models as a basis for pest control, p. 195-213. *In* J.N.R. Jeffers [ed.], Mathematical models in ecology, 12th Symp. Brit. Ecol. Soc. Blackwells, Oxford.

Conway, G.R., G.A. Norton, A.B.S. King, and N.J. Small. A systems approach to the control of the sugar cane froghopper. *In* G.E. Dalton [ed.], Agricultural systems. Applied Science, Barking, Essex. In press.

Crombie, A.C. 1945. On competition between different species of graminivorous insects. Proc. Roy. Soc. (B) 132:362-95.

———. 1946. Further experiments on insect competition. Proc. Roy. Soc. (B) 133:76-109.

Dempster, J.P. 1971. A population study of the cinnabar moth, *Tyria (Callimorpha) jacobaeae* L., p. 380-89. *In* P.J. den Boer and G.R. Gradwell [eds.], Dynamics of populations. Proc. Advan. Study Inst. Dyn. Numbers Pop. Oosterbeek, Netherlands, 1970. Pudoc, Wageningen.

———. Animal population ecology. Academic Press, London. In press.

Diamond, P. Stochastic stability of a host-parasite model. Maths. Biosci., in press.
Dingle, H. 1968. Life history and population consequences of density, photoperiod, and temperature in a migrant insect, the milkweed bug, *Oncopeltus*. Amer. Natur. 102:149-63.
———. 1972. Migration and strategies of insects. Science 175:1327-35.
———. 1974. The experimental analysis of migration and life history strategies in insects, p. 329-42. *In* L. Barton Browne [ed.], Experimental analysis of insect behaviour. Springer, New York.
Duffey, E. 1968. Ecological studies on the large copper butterfly *Lycaena dispar* Haw. *batavus* Obth. at Woodwalton Fen National Nature Reserve, Huntingdonshire. J. Appl. Ecol. 5:69-96.
Ehrlich, P.R., and L.E. Gilbert. 1973. Population structure and dynamics of the tropical butterfly *Heliconius ethilla*. Biotropica 5:69-82.
Elsey, K.D., and R.L. Rabb. 1970. Analysis of the seasonal mortality of the cabbage looper in North Carolina. Ann. Entomol. Soc. Amer. 63:1597-1604.
Embree, D.G. 1965. The population dynamics of the winter moth in Nova Scotia 1954-1962. Mem. Entomol. Soc. Can. 46:1-57.
Entwistle, P.F., C.G. Johnson, and E. Dunn. 1959. New pests of cocoa (*Theobroma cacao* L.) in Ghana following applications of insecticides. Nature, London 184:2040.
Fransz, H.G. 1974. The functional response to prey density in an acarine system. Pudoc, Wageningen.
Gilbert, N., and A.P. Gutierrez. 1973. A plant-aphid-parasite relationship. J. Anim. Ecol. 42:323-40.
Gilbert, N., and R.D. Hughes. 1971. A model of an aphid population—three adventures. J. Anim. Ecol. 40:525-34.
Gutierrez, A.P., P.E. Havenstein, H.A. Nix, and P.A. Moore. 1974. The ecology of *Aphis craccivora* Koch. and subterranean clover stunt virus in South-East Australia. II. A model of cowpea aphid populations in temperate pastures. J. Appl. Ecol. 11:1-20.
Haldane, J.B.S. 1953. Animal populations and their regulation. New Biol. 15:9-24.
Harcourt, D.G. 1971. Population dynamics of *Leptinotarsa decemlineata* (Say) in Eastern Ontario. III. Major population processes. Can. Entomol. 103:1049-61.
Hassell, M.P. 1966. Evaluation of parasite or predator response. J. Anim. Ecol. 35:65-75.
———. 1968. The behavioural response of a tachinid fly (*Cyzenis albicans* (Fall.)) to its host the winter moth

(*Operophtera brumata* (L.)). J. Anim. Ecol. 37:627-39.
———. 1971*a*. Mutual interference between searching insect parasites. J. Anim. Ecol. 40:473-86.
———. 1971*b*. Parasite behaviour as a factor contributing to the stability of insect host-parasite interactions, p. 366-79. *In* P.J. den Boer and G.R. Gradwell [eds.], Dynamics of populations. Proc. Advan. Study Inst. Dyn. Numbers Pop., Oosterbeek, Netherlands, 1970. Pudoc, Wageningen.
———. Density dependence in single-species populations. J. Anim. Ecol. 44, in press.
Hassell, M.P., and C.B. Huffaker. 1969. The appraisal of delayed and direct density dependence. Can. Entomol. 101: 353-61.
Hassell, M.P., and R.M. May. 1973. Stability in insect host-parasite models. J. Anim. Ecol. 42:693-719.
———. 1974. Aggregation of predators and insect parasites and its effect on stability. J. Anim. Ecol. 43:567-87.
Hassell, M.P., and D.J. Rogers. 1972. Insect parasite responses in the development of population models. J. Anim. Ecol. 41:661-76.
Hassell, M.P., and G.C. Varley. 1969. New inductive population model for insect parasites and its bearing on biological control. Nature, London 223:1133-37.
Holling, C.S. 1959. The components of predation as revealed by a study of small mammal predation on the European pine sawfly. Can. Entomol. 91:293-320.
———. 1961. Principles of insect predation. Annu. Rev. Entomol. 6:163-82.
———. 1964. The analysis of complex population processes. Can. Entomol. 96:335-47.
———. 1966. The functional response of invertebrate predators to prey density. Mem. Entomol. Soc. Can. 48: 3-86.
———. 1973. Resilience and stability of ecological systems. Annu. Rev. Ecol. Syst. 4:1-23.
Huffaker, C.B. 1958. Experimental studies on predation: dispersion factors and predator-prey oscillations. Hilgardia 27:343-83.
Huffaker, C.B., C.E. Kennett, and G.L. Finney. 1962. Biological control of olive scale *Parlatoria oleae* (Colvée), in California by imported *Aphytis maculicornis* (Masi)(Hymenoptera: Aphelinidae). Hilgardia 32:541-636.
Huffaker, C.B., and C.E. Kennett. 1966. Studies of two parasites of olive scale *Parlatoria oleae* (Colvée). IV. Biological control of *Parlatoria oleae* (Colvée) through

the compensatory action of two introduced parasites. Hilgardia 37:283-335.

Huffaker, C.B., C.E. Kennett, B. Matsumoto, and E.G. White. 1968. Some parameters in the role of enemies in the natural control of insect abundance, p. 59-75. *In* T.R.E. Southwood [ed.], Insect abundance. Roy. Entomol. Soc. London Symp. 4. Blackwell Scientific Publ., Oxford.

Huffaker, C.B., and P.S. Messenger. 1964. The concept of significance of natural control, p. 74-117. *In* P. DeBach [ed.], Biological control of insect pests and weeds. Chapman & Hall, London.

Hughes, R.D., and N. Gilbert. 1968. A model of an aphid population—a general statement. J. Anim. Ecol. 37:553-63.

Hughes, R.D., and B. Mitchell. 1960. The natural mortality of *Eroischia brassicae* (Bouché) (Dipt. Anthomyiidae): life tables and their interpretations. J. Anim. Ecol. 29: 359-74.

Ito, Y. 1961. Factors that affect the fluctuations of animal numbers, with special reference to insect outbreaks. Bull. Nat. Inst. Agr. Sci. Ser. C 13:57-89.

———. 1972. On the methods for determining density-dependence by means of regression. Oecologia (Berlin) 10:347-72.

Ito, Y., and K. Miyashita. 1968. Biology of *Hyphantria cunea* Drury (Lepidoptera: Arctiidae) in Japan. V. Preliminary life tables and mortality data in urban areas. Res. Pop. Ecol. 10:177-209.

Ito, Y., A. Shibazaki, and O. Iwahashi. 1969. Biology of *Hyphantria cunea* Drury (Lepidoptera: Arctiidae) in Japan. IX. Population dynamics. Res. Pop. Ecol. 11:211-28.

Iwao, S. 1971. Dynamics of numbers of a phytophagous lady-beetle, *Epilachna vigintioctomacubata*, living in patchily distributed habitats, p. 129-47. *In* P.J. den Boer and G.R. Gradwell [eds.], Dynamics of populations. Proc. Advan. Study Inst. Dyn. Numbers Pop., Oosterbeek, Netherlands, 1970. Pudoc, Wageningen.

Johnson, C.G. 1969. Migration and dispersal of insects by flight. Methuen, London.

Kiritani, K. 1964. Natural control of populations of the southern green stink bug, *Nezara viridula*. Res. Pop. Ecol. 6:88-98.

Kiritani, K., N. Hokyo, T. Sasaba, and F. Nakasuji. 1970. Studies on population dynamics of the green rice leafhopper, *Nephotettix cincticeps* Uhler: regulatory mechanism of the population density. Res. Pop. Ecol. 12:137-53.

Klomp, H. 1966*a*. The interrelations of some approaches to the

concept of density dependence in animal populations. Meded. Landbouwhogesch. Wageningen 66(3):1-11.
———. 1966*b*. The dynamics of a field population of the pine looper, *Bupalus piniarius* L. (Lep. Geun.). Advan. Ecol. Res. 3:207-305.
———. 1968. A seventeen year study of the abundance of the pine looper, *Bupalus piniarius* L. (Lepidoptera: Geometridae), p. 98-108. *In* T.R.E. Southwood [ed.], Insect abundance. Roy. Entomol. Soc. London Symp. 4. Blackwell Scientific Publ., Oxford.
Kokubo, A. 1965. Population fluctuations and natural mortalities of the pine-moth, *Dendrolimus spectabilis*. Res. Pop. Ecol. 7:23-34.
Kuno, E. 1971. Sampling error as a misleading artifact in "key factor" analysis. Res. Pop. Ecol. 13:28-45.
Landrahl, J.T., and R.B. Root. 1969. Differences in the life tables of tropical and temperate milkweed bugs, genus *Oncopeltus* (Hemiptera: Lygaeidae). Ecology 50:734-37.
Laughlin, R. 1965. Capacity for increase: a useful population statistic. J. Anim. Ecol. 34:77-91.
Lawton, J.H. 1974. Book review: animal interactions. Nature, London 248:537.
Lewontin, R.C. 1969. The meaning of stability. Brookhaven Symp. Biol. 22:13-24.
MacArthur, R.H. 1960. On the relative abundance of species. Amer. Natur. 94:25-36.
Maelzer, D.A. 1970. The regression of log N_{n+1} on log N_n as a test of density dependence: an exercise with computer-constructed density-independent populations. Ecology 51: 810-22.
Massee, A.M. 1952. Fluctuations in orchard fauna. Rep. Int. Hort. Congr. 13:1-5.
May, R.M. 1973. Stability and complexity in model ecosystems. Princeton University Press, Princeton.
———. 1974. Stability in ecosystems: some comments, p. 67. *In* Structure, functioning and management of ecosystems. Proc. 1st. Int. Congr. Ecol.
———. Biological populations obeying difference equations: stable points, stable cycles, and chaos. J. Theor. Biol., in press.
May, R.M., G.R. Conway, M.P. Hassell, and T.R.E. Southwood. 1974. Time delays, density-dependence and single-species oscillations. J. Anim. Ecol. 43:747-70.
Maynard Smith, J. 1974. Models in ecology. University Press, Cambridge.
McNeill, S. 1973. The dynamics of a population of *Leptopterna*

dolabrata (Heteroptera: Miridae) in relation to its food resources. J. Anim. Ecol. 42:495-507.

Messenger, P.S. 1964. Use of life tables in a bioclimatic study of an experimental aphid-braconid wasp host-parasite system. Ecology 45:119-31.

Metcalfe, J.R. 1972. An analysis of the population dynamics of the Jamaican sugar-cane pest *Saccharosydne saccharivore* (Westw.)(Hom., Delphacidae). Bull. Entomol. Res. 62:73-85.

Milne, A. 1957. Theories of natural control of insect populations. Cold Spring Harbor Symp. Quant. Biol. 22: 253-71.

Morris, R.F. 1959. Single-factor analysis in population dynamics. Ecology 40:580-88.

———. 1963. The dynamics of epidemic spruce budworm populations. Mem. Entomol. Soc. Can. 31:1-332.

———. 1964. The value of historical data in population research, with particular reference to *Hyphantria cunca* Drury. Can. Entomol. 96:356-68.

Mukerji, M.K. 1971. Major factors in survival of the immature stages of *Hylemya brassicae* (Diptera: Anthomyiidae) on cabbage. Can. Entomol. 103:717-28.

Murdie, G., and M.P. Hassell. 1973. Food distribution, searching success and predator-prey models, p. 87-101. *In* M.S. Bartlett [ed.], Mathematical theory of the dynamics of biological populations. Academic Press, London.

Murdoch, W.W. 1969. Switching in general predators: experiments on predator specificity and stability of prey populations. Ecol. Monogr. 39:335-54.

———. 1971. The developmental response of predators to changes in prey density. Ecology 52:132-37.

Murdoch, W.W., and J.R. Marks. 1973. Predation by coccinellid beetles: experiments on switching. Ecology 54:160-67.

Murdoch, W.W., and A. Oaten. Predation and population stability. Advan. Ecol. Res., in press.

Neilson, M.M., and R.F. Morris. 1964. The regulation of European spruce sawfly numbers in the Maritime Provinces of Canada from 1937-1963. Can. Entomol. 96:773-84.

Nicholson, A.J. 1954. An outline of the dynamics of animal populations. Aust. J. Zool. 2:9-65.

Nicholson, A.J., and V.A. Bailey. 1935. The balance of animal populations. Part 1. Proc. Zool. Soc. London 1935:551-98.

Orians, G.H. 1974. Diversity, stability and maturity in natural ecosystems, p. 64-65. *In* Structure, functioning and management of ecosystems. Proc. 1st. Int. Congr. Ecol.

Park, T. 1948. Experimental studies on interspecies competition 1. Competition between populations of the flour beetles *Tribolium confusum* Duval and *Tribolium castaneum* Herbst. Ecol. Monogr. 18:265-308.

———. 1954. Experimental studies on interspecific competition II. Temperature, humidity and competition in two species of *Tribolium*. Physiol. Zool. 27:177-238.

Pickles, A. 1936. Observations on the early larval mortality of certain species of *Diatraea* (Lepid., Pyralidae) under cane field conditions in Trinidad. Trop. Agr. 13:155-60.

Podoler, H. 1974. Analysis of the life tables for a host and parasite (*Plodia-Nemeritis*) ecosystem. J. Anim. Ecol. 43: 653-70.

Podoler, H.G., and D. Rogers. A new method for the identification of key factors from life table data. J. Anim. Ecol. 44, in press.

Potts, G.R. 1970. The effects of the use of herbicides in cereals on the feeding ecology of partridges. Proc. 10th Brit. Weed Contr. Conf. 1:299-302.

Potts, G.R., and G.P. Vickerman. 1974. Studies on the cereal ecosystem. Advan. Ecol. Res. 8:107-97.

Readshaw, J.L. 1965. A theory of phasmatid outbreak release. Aust. J. Zool. 13:475-90.

Richards, O.W. 1940. The biology of the small white butterfly (*Pieris rapae*), with special reference to the factor controlling abundance. J. Anim. Ecol. 9:243-88.

———. 1961. The theoretical and practical study of natural insect populations. Annu. Rev. Entomol. 6:147-62.

Richards, O.W., and T.R.E. Southwood. 1968. The abundance of insects: introduction, p. 1-7. *In* T.R.E. Southwood [ed.], Insect abundance. Roy. Entomol. Soc. London Symp. 4. Blackwell Scientific Publ., Oxford.

Richards, O.W., and N. Waloff. 1954. Studies on the biology and population dynamics of British grasshoppers. Anti-Locust Bull. 17:1-184.

———. 1961. A study of a natural population of *Phytodecta olivacea* (Forster)(Coleoptera, Chrysomeloidea). Phil. Trans. Roy. Soc. London 244:205-51.

Ripper, W.E. 1956. Effect of pesticides on balance of arthropod populations. Annu. Rev. Entomol. 1:403-48.

Rogers, D. 1972. Random search and insect population models. J. Anim. Ecol. 41:369-83.

Rogers, D.J., and M.P. Hassell. 1974. General models for insect parasite and predator searching behaviour: interference. J. Anim. Ecol. 43:239-53.

Royama, T. 1971. A comparative study of models for predation and parasitism. Res. Pop. Ecol. Suppl. 1:1-90.

St. Amant, J.L.S. 1970. The detection of regulation in

animal populations. Ecology 51:823-28.
Samson-Boshuizen, M., J.C. van Lenteren, and K. Bakker. 1974. Success of parasitization of *Pseudeucoila bochei* Weld (Hym., Cynip.): a matter of experience. Neth. J. Zool. 24:67-85.
Shoemaker, C. 1973*a*. Optimization of agricultural pest management I: biological and mathematical background. Math. Biosci. 16:143-75.
———. 1973*b*. Optimization of agricultural pest management II: formulation of a control model. Math. Biosci. 17: 357-65.
———. 1973*c*. Optimization of agricultural pest management III: results and extensions of a model. Math. Biosci. 18: 1-22.
Smith, R.F. 1970. Pesticides: their use and limitations in pest management, p. 103-18. *In* R.L. Rabb and F.E. Guthrie [eds.], Concepts of pest management. N.C. State University, Raleigh.
Solomon, M.E. 1949. The natural control of animal populations. J. Anim. Ecol. 18:1-35.
Southwood, T.R.E. 1960. The flight activity of Heteroptera. Trans. Roy. Entomol. Soc. London 112:173-220.
———. 1962*a*. Migration of terrestrial arthropods in relation to habitat. Biol. Rev. 37:171-214.
———. 1962*b*. Migration—an evolutionary necessity for denizens of temporary habitats. 11th Int. Kong. Entomol. 3:55-58.
———. 1966. Ecological methods. Methuen & Co., London.
———. 1967. The interpretation of population change. J. Anim. Ecol. 36:519-29.
———. 1968. The abundance of animals. Inaug. Lect. Imp. Coll. Sci. Technol. 8:1-16.
Southwood, T.R.E., and W.F. Jepson. 1962. Studies on the populations of *Oscinella frit* L. (Dipt. Chloropidae) in the oat crop. J. Anim. Ecol. 31:481-95.
Southwood, T.R.E., R.M. May, M.P. Hassell, and G.R. Conway. 1974. Ecological strategies and population parameters. Amer. Natur. 108:791-804.
Southwood, T.R.E., G. Murdie, M. Yasuno, R.J. Tonn, and P.M. Reader. 1972. Studies on the life budget of *Aedes aegypti* in Wat Samphaya, Bangkok, Thailand. Bull. World Health Organ. 46:211-26.
Spradbery, J.P. 1971. Seasonal changes in the population structure of wasp colonies (Hymenoptera: Vespidae). J. Anim. Ecol. 40:501-23.
———. 1973. Wasps. Sidgwick & Jackson, London.

Steele, J.H. 1974. The structure of marine ecosystems. Blackwell Scientific Publ., Oxford.
Stortenbecker, C.W. 1967. Observations on the population dynamics of the red locust, *Nomadacris septemfasciata* (Serville), in its outbreak areas. Meded. Inst. Toegep. Biol. Onderz. Natur. 84:1-118.
Sweet, M.H. 1963. The biology and ecology of the Rhyparochrominae of New England (Heteroptera: Lygaeidae). Part 1. Entomol. Amer. 43:1-124.
Takahashi, F. 1964. Reproduction curve with two equilibrium points: a consideration on the fluctuation of insect population. Res. Pop. Ecol. 6:28-36.
Utida, S. 1950. On the equilibrium state of the interacting population of an insect and its parasite. Ecology 31:165-75.
———. 1955. Fluctuations in the interacting populations of host and parasite in relation to the biotic potential of the host. Ecology 36:202-6.
Varley, G.C. 1963. The interpretation of change and stability in insect populations. Proc. Roy. Entomol. Soc. London (C) 27:52-57.
Varley, G.C., and G.R. Gradwell. 1960. Key factors in population studies. J. Anim. Ecol. 29:399-401.
———. 1968. Population models for the winter moth, p. 132-42. *In* T.R.E. Southwood [ed.], Insect abundance. Roy. Entomol. Soc. London Symp. 4. Blackwell Scientific Publ., Oxford.
———. 1970. Recent advances in insect population dynamics. Annu. Rev. Entomol. 15:1-24.
Varley, G.C., G.R. Gradwell, and M.P. Hassell. 1973. Insect population ecology: an analytical approach. Blackwell Scientific Publ., Oxford.
Voûte, A.D. 1946. Regulation of the density of the insect-populations in virgin-forests and cultivated woods. Arch. Néer. Zool. 7:435-70.
Walker, I. 1967. Effect of population density on the viability and fecundity in *Nasonia vitripennis*. Ecology 48: 294-301.
Waloff, N. 1968*a*. Studies on the insect fauna on Scotch Broom *Sarothamnus scoparius* (L.) Wimmer. Advan. Ecol. Res. 5:87-208.
———. 1968*b*. A comparison of factors affecting different insect species on the same host plant, p. 76-87. *In* T.R.E. Southwood [ed.], Insect abundance. Roy. Entomol. Soc. London Symp. 4.
Waloff, Z. 1966. The upsurges and recessions of the Desert

Locust plague: an historical survey. Anti-Locust Mem. 8.

Watmough, R.H. 1968. Population studies on two species of Psyllidae (Homoptera, Sternorhyncha) on broom (*Sarothamnus scoparius* (L.) Wimmer). J. Anim. Ecol. 32:283-314.

Watt, K.E.F. 1959. A mathematical model for the effect of densities of attacked and attacking species and the number attacked. Can. Entomol. 91:129-44.

———. 1963*a*. Mathematical population models for five agricultural crops. Mem. Entomol. Soc. Canada 32:84-91.

———. 1963*b*. Dynamic programming, "look ahead programming", and the strategy of insect pest control. Can. Entomol. 95: 525-36.

———. 1969. A comparative study on the meaning of stability in five biological systems: insect and fur beaver populations, influenza, Thai haemorrhagic fever and plague. Brookhaven Symp. Biol. 22:142-50.

Way, M.J. 1968. Intra-specific mechanisms with special reference to aphid populations, p. 18-36. *In* T.R.E. Southwood [ed.], Insect abundance. Roy. Entomol. Soc. London Symp. 4.

———. 1973. Objectives, methods and scope of integrated control, p. 138-152. *In* P.W. Geier, L.R. Clark, D.J. Anderson, and H.A. Nix [eds.], Insects: studies in population management. Ecol. Soc. Aust. (Mem. 1) Canberra.

Whittaker, R.H. 1974. "Stability" in plant communities, p. 68. *In* Structure, functioning and management of ecosystems. Proc. 1st Int. Congr. Ecol.

Whittaker, R.H., S.A. Levin, and R.B. Root. 1973. Niche, habitat and ecotope. Amer. Natur. 107:321-38.

Williamson, M. 1972. The analysis of biological populations. Edward Arnold, London.

Young, A.M., and A. Muyshondt. 1972. Biology of *Morpho polyphemus* (Lepidoptera: Morphidae) in El Salvador. J. N.Y. Entomol. Soc. 80:18-42.

PARASITES, PREDATORS,

AND POPULATION DYNAMICS

Powers S. Messenger

It is indeed a pleasure to share in the celebration of the centennial of entomology at Cornell University, and particularly to honor John Henry Comstock, who played such an important role in entomological education in North America. Both Professor Comstock and the Department of Entomology at Cornell have made their impact throughout the nation, and I can provide clear evidence that California has come under their influence. Cornell entomology alumni have appeared in all entomological quarters in the University of California and have held positions ranging from agricultural deans and associate deans to department chairmen, and they are now, and for long have been, numerous on the faculties at Berkeley, Davis, and Riverside, California.

Speaking about Professor Comstock's legacies to entomology in North America, perhaps I should mention a recent example, one which should interest all applied entomologists. This is the invasion of California by the famous Comstock mealybug, *Pseudococcus comstocki* (Kuwana). This pest of many crop plants is now in California's great Central Valley, the center of attention for all sorts of control endeavors, ranging from eradication attempts, to area-wide suppression with chemicals, to biological control with parasites and predators.

With Comstock's mealybug as a point of departure, I shall introduce my topic: parasites, predators, and population dynamics. Parasitoids and predators have interested applied entomologists for about as many years as entomology has been an academic subject at Cornell. Just 101 years ago (in 1873), C.V. Riley, then entomologist for the State of Missouri, initiated the first international shipment of a predator for purposes of controlling a crop pest when he shipped the predatory mite *Tyroglyphus phylloxerae* Riley to France to control the grape phylloxera. And ninety-one years ago the first exotic parasitoid, *Apanteles glomeratus* (L.), was introduced into North America from abroad for the control of the cabbage butterfly *Pieris rapae* (L.). Indeed, John Henry and Anna Botsford Comstock referred to some of these natural enemies in general terms, particularly some parasitic Hymenoptera and predatory Coccinellidae, in their *Manual of the Study of Insects*, first published at Cornell in 1895.

However, it is not my purpose to detail further the historical development of our knowledge of parasitoids and predators, but rather to summarize our current views of their role in influencing the population dynamics of other insects, and to examine to a limited extent their own population dynamics.

Population dynamics refers to the rising and falling patterns of abundance of populations over time, and the causes and limitations of such fluctuations. These fluctuations are not random, but rather are the result of causal mechanisms that reside within the environment of the given population. The occurrence of such contained population fluctuations, their numerical limitations, and the controlling mechanisms are included in a concept we call *natural control*. Hence *population dynamics* refers to the levels of and changes in density, over time, while *natural control* refers to the environmental mechanisms and factors whose actions determine and alter these densities.

Biological control, incidentally, is that aspect of natural control that results from the impact of natural enemies: parasitoids, predators, and pathogens. For present purposes we can distinguish the intentional use of natural enemies by man to control crop pests as applied biological control. I shall use the results of certain programs of applied biological control carried out in California to illustrate some of my points.

We frequently find that one of the major factors of the environment that enters into the natural control of insect abundance is natural enemies. This is not to say that all insect populations in nature are controlled numerically by natural enemies or even that in any given case a population is numerically limited only by natural enemies. It does mean that many such populations are affected by natural enemies, among other environmental factors, and also that in quite a few cases natural enemies are the main limiting or controlling factors.

Thus parasitoids and predators play important roles in more than just those crop-pest situations susceptible to applied biological-control attempts by man. The evidence for this contention is of two sorts, the results of numerous successful attempts at biological control of insect pests and weeds, and the unintended side effects of numerous programs of chemical control of insect pests of crops.

Before leaving this preliminary discussion of population dynamics, it should be mentioned that populations of natural enemies, particularly as the term refers to insect predators and parasitoids, are themselves subject to dynamical fluctuations, just as their hosts and prey are. However, the major contributory factor in the dynamics of predators and parasitoids is the dynamical changes in numbers of these host or prey populations. As is well known and accepted by ecologists, the density-related numerical interactions (density dependence) between host or prey on the one hand and parasitoid or predator on the other not only "control" the numbers of the former, but also "determine" the numbers of the latter. Hence we speak of and can demonstrate experimentally such paired population curves as host-parasite and predator-prey oscillations.

Having set the framework for an exploration of various features of predators, parasitoids, and population dynamics, we are now in a position to examine the following interesting ecological phenomena or concepts:

1. host-parasitoid interactions in the laboratory and field;
2. hyperparasitoids and their effects on host-parasitoid interactions;
3. competition between phytophagous insect species and the effects of natural enemies thereon;
4. species diversity, community stability, and agricultural monocultures, and the possibilities for biological control in such restricted ecosystems;
5. host specificity, parasitoid versus predator, and integrated control of insect pests of agricultural crops.

HOST-PARASITOID INTERACTIONS

The trophic interaction between a phytophagous host insect and a natural enemy can be a major cause of the population fluctuations observable in nature. Such fluctuations have, of course, since 1925 been predicted by theory (Lotka-Volterra cycles, Nicholson-Bailey oscillations) and demonstrated in laboratory experiments where the only realized limitations on host populations are natural enemies. Utida (1957) has shown such fluctuations over more than a hundred generations in the bean weevil, *Callosobruchus chinensis* (L.), and a pteromalid larval parasite, *Heterospilus prosopidis* Viereck (fig. 1). He has also shown

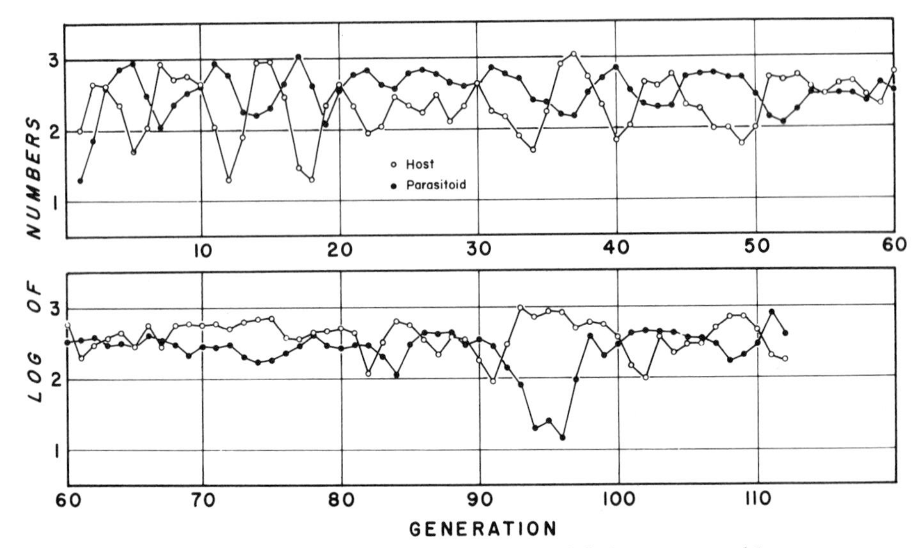

Fig. 1. Population cycles for the azuki bean weevil, Callosobruchus chinensis *(L.) (-O-), caused by the parasitoid* Heterospilus prospopidis *Viereck (-●-) in laboratory cultures. Censuses are total adults of each indicated species produced each generation. (Modified after Utida 1957.)*

that two parasitoids, in this case *H. prosopidis* and a braconid wasp, *Neocatolaccus mamezophagus* Ishii and Nagasawa, can coexist while controlling the same host in laboratory containers (petri dishes)(fig. 2). This was a situation which until recently was inexplicable from the standpoint of the two aforementioned theoretical models, the Lotka-Volterra and Nicholson-Bailey models of host-parasitoid interaction, and the general theory of competitive exclusion. I shall come back to this point later.

My colleague, C.B. Huffaker, and one of his graduate students, E.G. White, have demonstrated similar natural-enemy-induced population fluctuations in the Mediterranean flour moth, *Anagasta kühniella* (Zeller), caused by the ichneumonid larval parasitoid *Venturia canescens* (Gravenhorst)(White and Huffaker 1969*b*). These laboratory populations were sustained for more than three years, or twenty-two consecutive generations (fig. 3). Repeated host-population cycles four

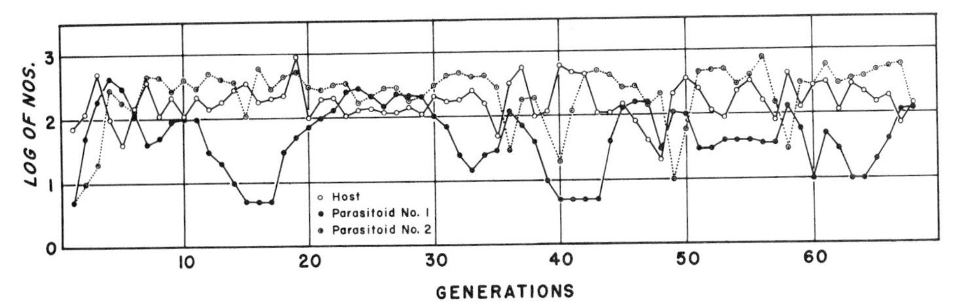

Fig. 2. Population cycles of the azuki bean weevil, Callosobruchus chinensis *(L.) (-O-), caused by the simultaneous attack of two parasitoids,* Heterospilus prosopidis *Viereck (-●-) and* Neocatolaccus mamezophagus *Ishii and Nagasawa (-⊙-), in laboratory cultures. Censuses are total adults of each indicated species produced each generation. (Modified after Utida 1957.)*

generations long occurred in this case. In control experiments, where parasitoids were absent, food became the limiting factor, and no cyclicity in the host population occurred.

The controlling natural enemy need not be a parasitoid. Predators may also play this role. White and Huffaker (1969*a*) show that cyclical host-enemy interactions with the same host, *A. kühniella*, can result from the action of the acarine egg predator, *Blattisocius tarsalis* (Berlese). In this case (fig. 4) the host fluctuates in a series of "decaying oscillations" that gradually die out, only to start up suddenly in an "outbreak" situation where food, and intraspecific competition for food, rather than the predator, are the limiting factors. These "outbreak and decay" cycles occur with a periodicity of about one year, seven host generations.

Thus natural enemies (predators and primary parasitoids) can act to control and bring about fluctuations in their prey or host populations. *Control* here is used in the sense of reducing host-population densities to levels below which food (i.e., in other situations the host plant) is the limiting factor. It is this capability, of course, that is sought in pest-control programs involving biological control.

A recent, successful case of biological control in

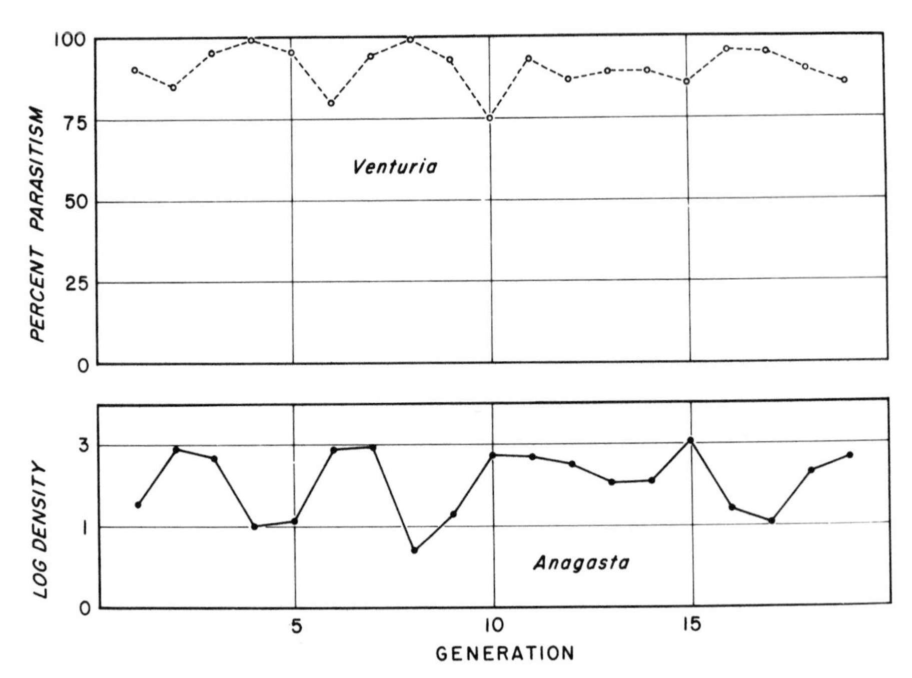

Fig. 3. Population cycles of the Mediterranean flour moth, Anagasta kühniella *(Zeller)(-●-), caused by the parasitoid* Venturia canescens *(Gravenhorst) in laboratory cultures, together with percent parasitism of immature host stages (-O-). Censuses are total moth adults produced each generation. (Modified after White and Huffaker 1969*b.*)*

California can serve to explain several interesting aspects of the impact of natural enemies on insect-population dynamics. This case involves the walnut aphid, *Chromaphis juglandicola* (Kaltenbach).

The walnut, *Juglans regia* (L.), is of course an exotic plant to North America, coming initially from Iran and the region of the Himalayan foothills in Asia. In California the principal insect pests of walnut are the codling moth, the walnut husk fly, and the walnut aphid. There are also a number of lesser pests, including mites and the navel orangeworm. In certain areas, as much a consequence of tree variety as climate, neither the codling moth nor the husk fly are pestiferous, and the main pest is the aphid.

The aphid is quite injurious and, unless checked by

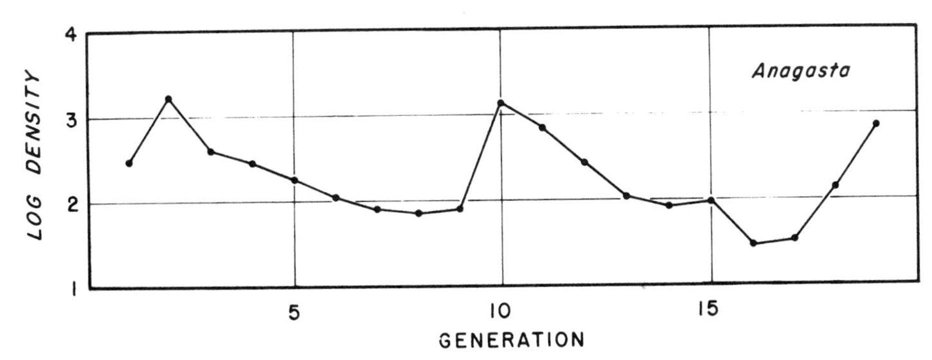

Fig. 4. Population cycles of the Mediterranean flour moth, Anagasta kühniella *(Zeller)(-●-), caused by the egg-consuming predatory mite,* Blattisocius tarsalis *(Berlese), in laboratory cultures. Censuses are total moth adults produced each generation. (Modified after White and Huffaker 1969*a.*)*

positive control measures, can result in loss of yield, reduced quality of the yield, production of sooty molds via the copious quantities of honeydew produced, and decline in vigor of the trees.

In California the aphid was but rarely attacked by parasitoids. On the other hand, it drew the interest of numerous predatory insects, but not in such a way as to preclude damage. The usual aphid predators were involved: coccinellids, lacewings, syrphids, anthocorids (Sluss 1967).

In considering the possible biological control of the walnut aphid, one immediately thinks of the "vacant niche" for a parasitoid that seems to exist in this aphid system. In late 1968 a parasitoid, *Trioxys pallidus* (Haliday)(Iranian ecotype), was introduced into northern and central California, and within three years gave completely successful control of the aphid pest (Van den Bosch et al. 1970; Van den Bosch and Messenger 1973). Schematic population curves in figure 5 give the "before and after" picture of this program.

The walnut aphid, a native of Asia just like its host, is apparently monophagous in North America, attacking no other plant but *Juglans regia*. The primary parasitoid, *T. pallidus*, is also an exotic in North America. As far as we know, it, too, is monophagous, attacking only the walnut aphid. Thus the entire interaction that has brought about the result described in figure 5 is contained in the trophic system of walnut, *Chromaphis*, and *T. pallidus*.

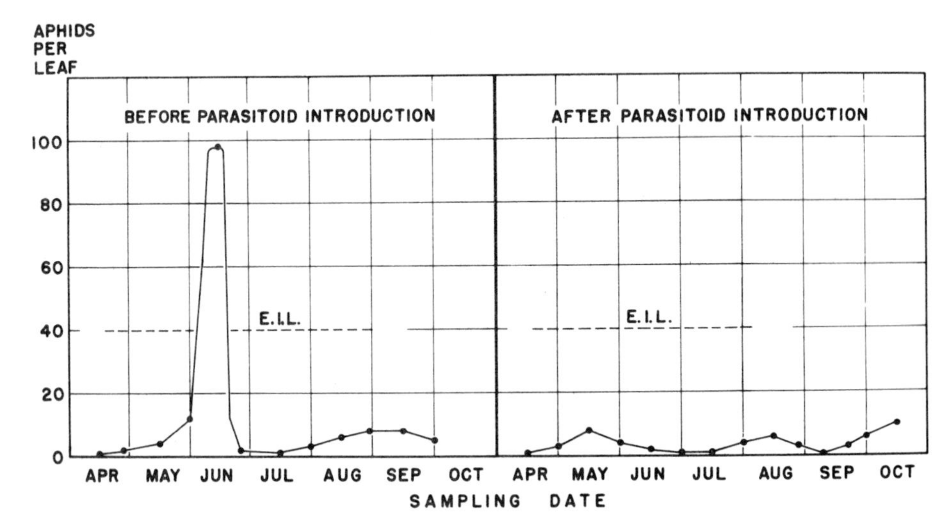

Fig. 5. Seasonal population curves for the walnut aphid, Chromaphis juglandicola *(Kaltenbach), before and after introduction of the Iranian ecotype of the parasitoid* Trioxys pallidus *(Haliday) in central California in 1969.* E.I.L. *shows the economic injury level of this pest on walnut. (Data for the former curve from Sluss 1967; for the latter curve, unpublished data from R. Hom, R. van den Bosch, and P.S. Messenger).*

This parasite must have a very efficient host-searching capacity, since in most periods of the growing season the aphid is now quite hard to find. The parasite seems to be beautifully synchronized with the host and host plant. The tree, of course, is deciduous. The aphid overwinters as a diapausing egg on twigs, leaf buds, and bark. The parasite overwinters as a diapausing prepupa within the aphid mummy attached to leaf buds, twigs, or dried leaves on the ground. Each spring, just when the leaf buds begin to swell and open, the aphid eggs hatch and the fundatrices begin developing. At the same time the parasite adults emerge from the mummy, mate, and promptly begin searching for and attacking aphids. Undoubtedly, this early parasitism of the aphid at the very onset of the season each year contributes to effective biological control.

Each of these events—initiation of leaf development, hatching of the aphid egg, and emergence of the parasite adult—is under the control of climate. The precise concordance of these three phenological events represents an extraordinary product of coevolution, for, clearly, if the aphid were to hatch earlier than the unfolding of the leaf, or the parasite emerge much before the aphid, serious populational injury would result.

HYPERPARASITOIDS

Soon after the parasitoid became established in central California, it was found to be attacked by hyperparasitoids. Every care was taken, of course, to prevent the importation of such noxious species during the biological-control program. We deduce that these hyperparasitoids came from the local California entomofauna, either being indigenous or accidentally introduced by man in perhaps the same way as were the various exotic-aphid pests in prior decades. Also, we can be sure that such secondary natural enemies were not associated with the walnut aphid prior to the introduction of *T. pallidus*. Suffice it to say that in the years 1969 to 1972 the frequency of hyperparasitism increased substantially, often reaching 80-90%, until now in late season nearly 100% of the primary parasitoids are themselves parasitized by secondaries.

We have been watching this situation carefully to see whether the degree of biological control is diminishing, and whether any other harmful influence on the primary can be detected. At least five different hyperparasitic species are involved; some of these may have rather narrow host ranges, and others are nonspecific.[1]

These hyperparasitoids belong to the genera *Asaphes*, *Pachyneuron*, *Dendrocerus* (=*Lygocerus*), *Aphidencyrtus*, and *Alloxysta* (=*Charips*). Careful observations by one of our graduate students[2] suggests that the season-end build-up of the hyperparasitoids may actually not be detrimental to biological control; but rather that they collectively may serve as a dampening agent, preventing the very efficient *T. pallidus* from overexploiting its host population. Parasitism rates of the early spring brood of aphids reach levels as high as 98% (figs. 6, 7, 8) even with very high hyperparasitism rates the previous autumn. This suggests that in the absence of such mortality the overwintering primary parasitoid population might annihilate its host resource, at least

---

[1] P. Matteson: unpublished data.

[2] B.D. Frazer: personal communication.

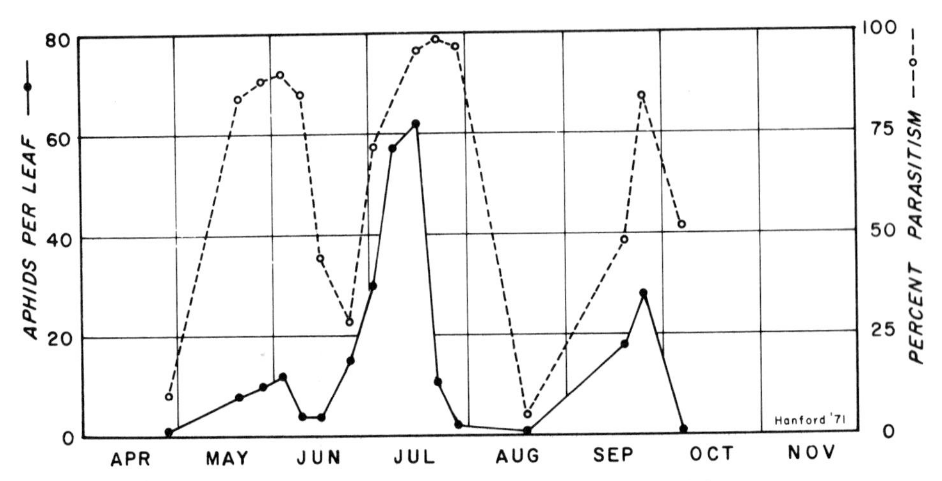

Fig. 6. Population curve for the walnut aphid, Chromaphis juglandicola *(Kalt.), and percent parasitism by the introduced parasitoid* Trioxys pallidus *(Hal.) in a commercial walnut orchard in the interior valley site, Hanford, San Joaquin Valley, California, 1971. (Unpublished data from R. Hom, R. van den Bosch, and P.S. Messenger.)*

in a local sense, resulting in the demise of the primary parasitoid itself, and a subsequent rapid, unchecked resurgence of the aphid (all adults being alate in this species). We are attempting to devise a method to test this hypothesis.

The various theories of host-parasitoid interaction provide no role for hyperparasitism. As Utida's studies suggest, neither do these conventional theories (Lotka-Volterra, Nicholson-Bailey, Gause's competitive exclusion principle) explain two coexisting parasitoids on the same host population. Of the two conventional theories for host-parasitoid interaction, the Lotka-Volterra model and the Nicholson-Bailey model, the latter is perhaps the more realistic in that it is based upon organisms having discrete generations and age-specific mortality relationships.

Recent work by Hassell and Huffaker (1969) has resulted in a simple modification of the Nicholson-Bailey host-parasitoid model, the incorporation of a variable "area of discovery" for the parasitoid, rather than a constant, as the parent model specifies. Empirical evidence shows clearly that the area of discovery of a parasitoid varies systematically (and inversely) with the density of the natural enemy. This has led to the Quest theory of Hassell and Varley (1969) and the idea of mutual interference (intraspecific competition) among parasitoids, which serves as a stabilizing factor for the host-parasitoid interaction. Indeed, with this Quest theory, Hassell and Varley are able to show that not only can the Nicholson-Bailey model produce stable oscillations of host and parasitoid, but two parasitoids may coexist while attacking the same host population. This is, of course, what Utida has demonstrated experimentally (fig. 2).

It is possible that a similar modification of the Nicholson-Bailey model that includes the dampening effects of a hyperparasite can also be devised. One way such an extrinsic dampening factor could operate is by superimposing a density-dependent mortality on the primary parasitoid in much the same fashion as mutual interference (intraspecific competition) does in the Quest theory of Hassell and Varley. That hyperparasitic insects may act as delayed density-dependent mortality factors, and thus will be "destabilizing," will have to be considered. However, the fact that many hyperparasitoids are not host specific—i.e., are general predators—may allow them to act more like opportunists, attacking the primary parasitoid when it becomes abundant and

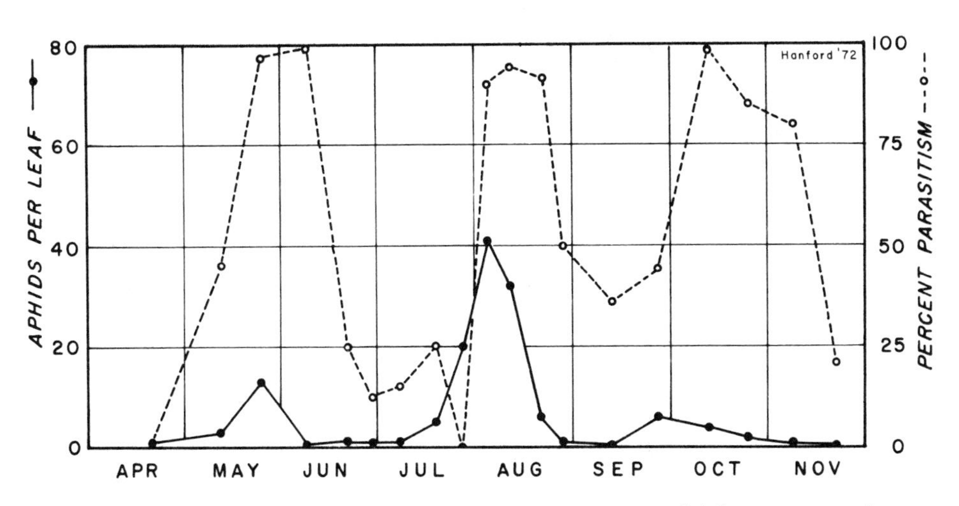

Fig. 7. Population curve for the walnut aphid, Chromaphis juglandicola *(Kalt.), and percent parasitism by the parasitoid* Trioxys pallidus *(Hal.), Hanford, California, 1972. (Unpublished data from R. Hom, R. van den Bosch, and P.S. Messenger.)*

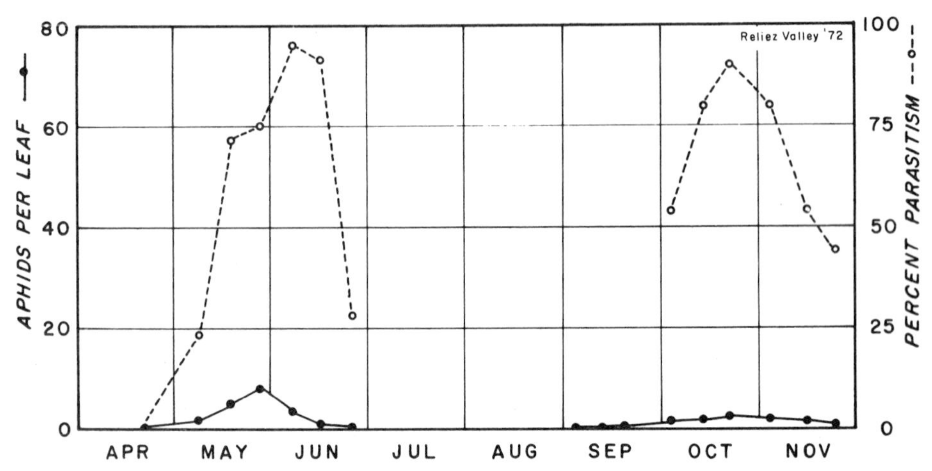

Fig. 8. Population curve for the walnut aphid, Chromaphis juglandicola *(Kalt.), and percent parasitism by the parasitoid* Trioxys pallidus *(Hal.) in a commercial walnut orchard in the coastal mountain location, Reliez Valley, Contra Costa County, California, 1972. (Unpublished data from R. Hom, R. van den Bosch, and P.S. Messenger.)*

ignoring it when it is scarce. This would allow the secondary parasitoids to act as direct density-dependent mortality agents and thus be more amenable to inclusion in the Quest theory.

PEST COMPETITORS AND NATURAL ENEMIES

Competition has for long captured the interest of ecologists, and hardly any course in or text on ecology omits discussion of competition, competitive exclusion, niche overlap, ecological homologues, and so on. Pest-control specialists have probably wondered whether competition plays a role, one way or another, in setting the environmental limits to the abundance of a potential pest.

I suppose this has even come under consideration by biological-control specialists. The walnut-aphid biological-control program in California, which, as mentioned above, was begun in 1968 and reached full economic success by 1972, provides us with a remarkable example of what I call "competitive release." It also provides certain other unexpected species interactions.

Consider again the major pests of walnut in central California. Besides the walnut aphid, there is the codling moth, the walnut husk fly, and several other minor pests. One such minor pest is frosted scale, *Lecanium pruinosum* Coquillet, which only becomes pestiferous when DDT is used in walnut pest-control programs (Michelbacher, Swanson, and Middlekauff 1946; Bartlett and Ortega 1952). Another such minor walnut pest is the dusky-veined aphid, *Panaphis juglandis* (Goeze). This aphid is normally a rarity on walnut, and in fact does not appear in most lists of pests of this crop. We now find, after suppression of the walnut aphid, *C. juglandicola*, with the parasitoid *T. pallidus*, that the dusky-veined aphid is increasing to high densities in a number of localities, and is everywhere in central California more abundant than ever before. In a few orchards, growers have even felt the need to control it.

It is now our hypothesis that the dusky-veined aphid was kept in check by *C. juglandicola*. The latter species inhabits the underside of the walnut leaves; dusky-veined aphid, the upper surface. Honeydew, excreted by the walnut aphid, falls on and coats the upper surface of the leaves, which results in injury to dusky-veined aphid (Olson 1974). Dusky-veined aphid's own honeydew adds to the problem. The walnut aphid simply "sits under the umbrella," out of the rain of honeydew. So, when *C. juglandicola* is removed from the scene, or, more precisely, reduced to low numbers, this competitive interaction is ended, and the dusky-veined aphid increases in abundance (fig. 9).

However, this is not the only possible reason for the irruptions of dusky-veined aphid. Another hypothesis is that, with the disappearance of large populations of walnut aphid, and the copious quantities of honeydew that accompany them, aphid predators such as lacewings, coccinellids, syrphids, and reduviids (Sluss 1967) are either not attracted to walnut trees or do not reproduce thereon if they do colonize walnut orchards each spring. This reduction in the aphid predators favors dusky-veined aphid, thus accounting for its increase. There is also the possibility, mentioned by Olson (1974), that the elimination of chemical controls for walnut aphid has enabled *P. juglandis* to increase in numbers.

One final comment about natural enemies and the population dynamics of walnut pests concerns the walnut husk fly, *Rhagoletis completa* (Cresson). This is a recent pest of walnut in northern and central California, having invaded

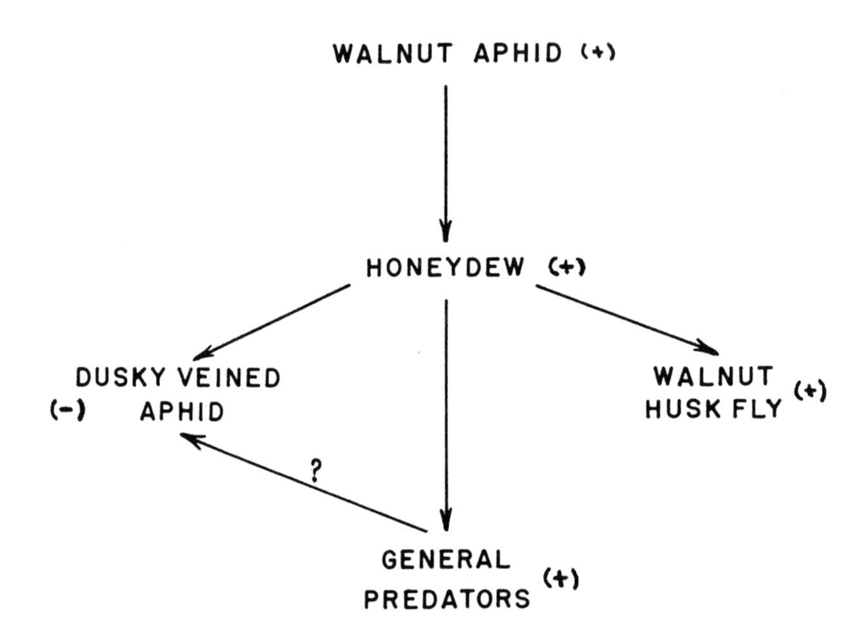

Fig. 9. Schematic diagram of the interaction between the walnut aphid, Chromaphis juglandicola *(Kalt.), and the associated species, the dusky-veined aphid,* Panaphis juglandis *(Goeze), and walnut husk fly,* Rhagoletis completa *(Cresson), and unspecified honeydew-eating aphid predators (see text). When the walnut aphid is abundant, signified by a (+) sign, other species or factors with the same sign are also abundant.*

this area in 1954. Prior to this time, *R. completa* was known only in southern California. The adult fly, on emergence from the puparium, must spend a period of one to two weeks feeding on proteinaceous and caloric foodstuffs. Honeydew is one such food. In a few cases in central California subsequent to the biological control of the walnut aphid, husk fly damage declined sharply. My colleague K.S. Hagen concludes that husk fly females are not attracted to walnut trees without honeydew, and hence they are not attracted to walnut orchards, nor do they oviposit as frequently on walnut fruit (fig. 9).

Host-enemy interaction, as we can see from the case of biological control of walnut aphid, can result in a number of strange and unexpected consequences.

COMPETING NATURAL ENEMIES

Not only may populations of different pest species compete with one another, but so also may different natural-enemy species attacking a common host. The consequences of such competition for both the host and the enemies themselves are various. A number of different examples are provided in programs of biological control.

It is common to describe competing species as either superior or inferior, according to which species is the "winner" and which the "loser." It is also customary to distinguish the site of competitive contact between two species as "intrinsic" or "extrinsic," according to whether the competitive interaction between natural-enemy individuals takes place within the host body, or between adult enemies away from the host (i.e., searching power, reproductive rate, longevity, or physical-factor tolerance range).

I will not examine further the intrinsic or extrinsic nature of competition between natural-enemy species, but will rather discuss the consequences of competition where the superiority or inferiority of a competitor is self evident from the results of natural-enemy introductions.

We can distinguish three different outcomes that may result from the introduction of a new natural-enemy species into an environment where one or more other natural enemies already attacks the target host. First, the colonization may fail because of the superiority of the preexisting enemy or enemies (competitive exclusion); second, there may be a replacement of a preexisting species by the new one (competitive displacement); third, the new enemy species may persist with those already present (coexistence).

In the first case described above, we rarely know that such is the actual reason for the failure, because it is a difficult and time-consuming task to verify that competitive exclusion, or rejection, has been the reason for the failure. It would appear to be well worth the effort to find that such has been the cause, but more often, for pragmatic reasons, biological-control specialists prefer to concentrate their efforts on their performance of natural-enemy species that do become established.

Competitive displacement, on the other hand, has been observed in a number of cases of biological control. In Hawaii, in the biological-control campaign against the oriental fruit fly, *Dacus dorsalis* Hendel, a sequence of parasitoids was colonized, involving in order the braconids *Opius longicandatus*, *O. vandenboschi*, and *O. oophilus* (Van den Bosch and Messenger 1973). Each became successfully established, but the first was displaced by the second, and the second by the third. At each step in the sequence of displacements, parasitism rates rose, and the average host density declined.

A similar sequence of displacements has occurred in respect to the California red scale in southern California citrus orchards (DeBach 1965). Here, the first established parasitoid, *Aphytis chrysomphalus* (Mercet), was displaced by *A. lingnanensis* Compere, which in turn was replaced over most of its range by *A. melinus* DeBach. Again, at each step in this sequence of displacements, the impact on host populations increased; i.e., biological control was improved.

Natural-enemy supplementation, the third possible result, is less common. We have already discussed the coexistence of Utida's two bean weevil parasitoids in the laboratory. One outstanding example from the field is the biological-control effort in northern and central California against the olive scale *Parlatoria oleae* (Colvée). In this case the first scale parasitoid introduced was *Aphytis maculicornis* (Masi). It reduced olive-scale densities substantially, providing effective control in certain locations and at certain times. This parasitoid suffered from the hot, dry summers characteristic of this part of California. Several years later a second enemy, *Coccophagoides utilis* Doutt, was colonized and became established also. Biological control thereafter was outstanding, obviating any need for other control measures against this pest. But rather than displacing *A. maculicornis*, *C. utilis* added its effects to those of the former. *A. maculicornis* remains active and dominant in the autumn, winter, and spring, while *C. utilis* is effective during the summer. When *A. maculicornis* is favored by climate, it is the superior competitor. In summer, when it becomes suppressed, *C. utilis* is able to evade the forces of competition and assist in inflicting mortality on the scale host (Huffaker and Kennett 1966).

DIVERSITY, STABILITY, AND MONOCULTURES

Many serious pest problems occur in crops grown on a large scale in massive blocks, as in many areas of the middle and western United States. In the Central Valley of California, where climate is favorable, water for irrigation ample, and the topography flat, a considerable number of crops are grown in what can be described literally as monocultures. Such crops include cotton, grape, alfalfa, safflower, and sugar beet. Seemingly to enhance this monoculture approach to modern agriculture, growers of such crops often practice intensive weed control along field boundaries, ditches, fences, and roadsides.

In situations as described above, floral diversity is reduced to the barest minimum. The question is, what effect does this floral impoverishment have on the occurrence and solution of insect-pest problems?

In recent years there has been considerable discussion, pro and con, about the concept that, in ecosystems, the more diverse the assemblage of species, the greater is the stability of the community. The concept is intuitively logical. The numerous insect-pest species, each in large numbers, in apparently simple agroecosystems such as those described above, lend support to the concept; so do the occasional flare-ups of certain pest populations after pesticide application, the reason presumably being the simplification of the community through destruction of natural enemies of such pests and possibly also the temporary release of some pest species from competition.

Biological control, of course, can be considered a method for increasing the species diversity of an agroecosystem, though it may occasionally do the opposite (see my example, above, concerning the walnut aphid). In those cases where biological control is successful in reducing phytophagous pest abundance, such increasing diversity can be said to have led to increased stability. Southwood and Way (1970) have cautioned that mere increase in species diversity, presumably leading to increasing linkages of the community foodweb, is not sufficient to provide stability. What must in addition be considered is the nature of the linkage: whether it is density-related, and more particularly whether it includes density-dependent regulative properties vis-à-vis the abundance of either or both of the species populations composing the linkage.

My colleague Carl Huffaker has recently considered this question of species diversity and community stability and cautions that these two aspects may not be cause and effect, but rather may both be the consequence of climatic stability (Huffaker 1974). Artificial modification of the one (diversity) may not necessarily alter the other (stability). On the other hand, there is a substantial body of empirical evidence showing improvements in crop-pest control in these monocultural ecosystems through man-made alterations in diversity.

Biological-control agents often possess the density-dependent properties required to provide the stabilizing capacity specified by Southwood and Way. Besides the introduction of new natural enemies into a crop-pest situation, as I have already described with regard to the walnut aphid in central California, biological-control specialists have attempted to augment natural-enemy activity in a variety of ways that might be described as efforts to increase diversity in order to achieve stability (of the pest).

The classic example is provided by R.L. Doutt and colleagues, who improved the performance of the native egg parasitoid, *Anagrus epos* Girault. This minute wasp, which is associated with the grape leafhopper, *Erythroneura elegantula* (Osborn), a key pest of grapevines in central California, is reduced in abundance each year in commercial vineyards through failure to overwinter. Its pest host, *E. elegantula*, overwinters as the adult, whereas the necessary host stage for the parasitoid is the egg. After careful evaluation, Doutt found that the preferred host of this parasitoid is another leafhopper which does overwinter in the egg stage, but which lives on an altogether different host plant; namely, blackberry, *Rubus* spp. Under normal cultivation practice, little or no *Rubus* brambles are allowed to grow in the very large (monoculture) vineyards of the Central Valley of California. By "interplanting" hedgerows of blackberry near vineyards, the parasitoid is substantially increased in abundance, particularly early in the season, a result which then provides substantially better control of the grape leafhopper than before (Doutt and Nakata 1965).

Another approach has been the addition of shelters to treeless and hedgeless crop fields for such natural enemies as predatory wasps (Lawson et al. 1961; Rabb 1971), and coccinellids.[1] The addition of artificial foodstuffs used by certain natural enemies, *Chrysopa* spp. and coccinellids, for example (Hagen and Tassan 1970; Hagen, Sawall, and Tassan 1970), can be considered an augmentation of diversity of the crop community; it increases the abundance of these predators and leads to reduced populations of pest species. The predators would normally get such foods from weeds and other plants, and the noneconomic insect species that feed on them,

---

[1]K.S. Hagen: unpublished data.

and thus the added foodstuffs can be considered as practical substitutes for such plants and herbivores.

The strip-cropping technique proposed for harvesting alfalfa by my colleagues in California (Schlinger and Dietrick 1960; Stern, Van den Bosch, and Leigh 1964) is a method for maintaining the diversity of an agroecosystem. In such fields alternate portions of the alfalfa crop are retained with their associated herbivorous, as well as carnivorous, insect populations. This promotes or maintains the biological control of alfalfa pests. Solid-cropping of alfalfa, the conventional harvesting procedure, periodically destroys local populations of insects and any balance that exists between herbivore and natural enemy.

PARASITOIDS VERSUS PREDATORS

Although several predators have been the agents providing successful biological control of crop pests, most successful cases of biological control are due to parasitoids (Doutt and DeBach 1964; Van den Bosch and Messenger 1973). However, we know that some predators, alone, can regulate the abundance of their prey (see discussion, above, of the control of *A. kühniella* by the egg predator, *Blattisocius tarsalis*; there is also, of course, the famous Vedalia lady beetle which controls the cottony cushion scale). The complex of general predators plays an important role in such field crops as cotton, alfalfa, maize, and sorghum.

I have mentioned the agronomic practice of monoculture of numerous crops in California and elsewhere in the western United States. When such planting practices are combined with intensive weed suppression, the agroecosystems appear to be excessively "simplified." One wonders whether much of a higher trophic structure (herbivores, carnivores) can exist, aside from the few obvious insect pests.

However, the simplicity of such monocultures is really more apparent than real. My colleagues R. van den Bosch, K.S. Hagen, and E.T. Schlinger have carried out at one time or another over the past fifteen to twenty years a number of intensive samplings of the arthropod fauna of both alfalfa and cotton fields in California. The collective number of arthropods found associated with either crop is 500-600. In any one field the number of species may reach as high as 150-250. By far the greater proportion of these are noneconomic. Probably only 6-10 species are pests, though some 15-20 can be classed as potential pests.

In this complex of arthropods, which includes insects, mites, spiders, centipedes, millipedes, isopods, and so on, there are members of many trophic levels: plant feeders, detritus feeders, decomposers, predators, parasitoids, and hyperparasitoids. Most of these are held in numerical check by others in the system. Many reside on or in the plant, others on the soil surface, still others in the soil.

In many such agroecosystems—cotton, alfalfa, safflower, and sorghum in particular—some of the most important arthropods are the general predators, such as *Chrysopa carnea* Stephens, *Nabis* spp., coccinellids, *Geocoris* spp., *Zelus* spp., *Orius* spp. (minute pirate bugs), and lycosid spiders. These ubiquitous carnivores attack most other soft-bodied or less pugnacious arthropods, particularly the eggs and small larvae of such other groups as Lepidoptera, Coleoptera, Hemiptera, and Homoptera. It is probably this group of roving marauders that is responsible for the fact that so many of the phytophagous arthropods in such crop habitats are not abundant enough to be harmful to crop production (Ehler and Van den Bosch 1974).

One good example of the importance of these insect predators comes from the work of University of California entomologists in their successful development of an integrated control program for cotton. Some of the most important *potential* pests of cotton are these familiar species: bollworm (*Heliothis zea* [Boddie]), armyworms (*Spodoptera* spp.), cabbage looper (*Trichophisia ni*), *Pseudaletia* spp., *Bucculatrix* spp., and *Laphygma* spp. These species, mostly native and mostly not kept in check by native or introduced parasitoids, are normally suppressed by the above-mentioned general predators. Use of a pesticide, however, destroys these natural enemies even though such materials may be effective against one or more pests; and these other, potential pests soon become real contenders for the growers' crops. The California cotton integrated control program has been a primarily successful effort to devise means for managing certain key pests, such as lygus bug, *Lygus hesperus*, and, where it occurs, pink bollworm, *Pectinophora gossypiella* (Saunders), without upsetting the populations of predatory natural enemies (Falcon et al. 1971).

The lesson to be learned from all this is that the apparent simplicity of monocultures (of plant species) is deceptive. They in fact contain a species complex that includes members with the trophic characteristics and numerical responsiveness appropriate for the maintenance of ecosystem stability.

The further lesson to be learned is that natural enemies are frequently the central elements in integrated-control schemes, and that, within the complex of natural enemies, those we describe as general predators are very frequently the principal actors.

SUMMARY

The abundance of phytophagous insect populations rises and falls over time according to environmental circumstances. Parasitoids and predators play roles in the population dynamics of such populations. Laboratory studies show that stable-population fluctuations can be produced in host species by even single natural-enemy species working alone.

Until recently, population theory has held that two or more natural enemies cannot coexist when attacking the same host population; one enemy species will, through competition, displace the other. There is a considerable body of empirical evidence that supports this theoretical conclusion. However, there are also several cases showing that natural enemies attacking the same host population can coexist. Recently, theoretical models have been modified to allow for this outcome.

Populations of insects interact in various ways, either directly or indirectly, with other insect populations. The biological control of the walnut aphid in California gives an example of how a natural enemy can suppress a phytophagous insect species, how insect hyperparasites can enter into the interaction, perhaps in a positive way with respect to pest suppression, and how suppression of one phytophagous insect population may allow another unrelated insect population to increase in abundance. Further "symbiotic" associations are upset or modified by such occurrences.

Agricultural ecosystems are usually looked on as overly simplified communities, with consequent instabilities in terms of population dynamics of associated arthropods. Agri-monocultures would appear to carry this to the extreme. Many pest-control attempts have been based on the artificial increase in this low-level diversity in hopes that stability of arthropod populations may result. General insect predators play a strong role in potential pest suppression, even in monocultures, and represent key factors around which successful integrated-control programs can be constructed.

REFERENCES

Bartlett, B., and J.C. Ortega. 1952. Relation between natural enemies and DDT-induced increases in frosted scales and other pests of walnut. J. Econ. Entomol. 45:783-84.

DeBach, P. 1965. Some biological and ecological phenomena associated with colonizing entomophagous insects, p. 287-306. *In* H.G. Baker and G.L. Stebbins [eds.], Genetics of colonizing species. Academic Press, New York.

Doutt, R.L., and P. DeBach. 1964. Some biological control concepts and questions, p. 118-42. *In* P. DeBach [ed.], Biological control of insect pests and weeds. Chapman and Hall Ltd., New York and London.

Doutt, R.L., and J. Nakata. 1965. Overwintering refuge of *Anagrus epos* (Hymenoptera: Mymaridae). J. Econ. Entomol. 58:586.

Ehler, L.E., and R. van den Bosch. 1974. An analysis of the natural biological control of *Trichoplusia ni* (Lepidoptera, Noctuidae) on cotton in California. Can. Entomol. 106:1067-73.

Falcon, L.A., R. van den Bosch, J. Gallegher, and A. Davidson. 1971. Investigation of the pest status of *Lygus hesperus* in cotton in central California. J. Econ. Entomol. 64:56-61.

Hagen, K.S., E.F. Sawall, Jr., and R.L. Tassan. 1970. The use of food sprays to increase the effectiveness of entomophagous insects. Proc. Tall Timbers Conf. Ecol. Anim. Contr. Habitat Manag. 2:59-81.

Hagen, K.S., and R.L. Tassan. 1970. The influence of food Wheast® and related *Saccharomyces fragilis* yeast products on the fecundity of *Chrysopa carnea* (Neuroptera: Chrysopidae). Can. Entomol. 102:806-11.

Hassell, M.P., and C.B. Huffaker. 1969. Regulatory processes and population cyclicity in laboratory populations of *Anagasta kühniella* (Zeller)(Lepidoptera: Phycitidae). III. Development of population models. Res. Pop. Ecol. 11:186-210.

Hassell, M.P., and G.C. Varley. 1969. New inductive population model for insect parasites and its bearing on biological control. Nature 223:1133-37.

Huffaker, C.B. 1974. Some implications of plant-arthropod and higher-level, arthropod-arthropod food links. Environ. Entomol. 3:1-9.

Huffaker, C.B., and C.E. Kennett. 1966. Studies of two parasites of olive scale, *Parlatoria oleae* (Colvée). IV. Biological control of *Parlatoria oleae* (Colvée) through the compensatory action of two introduced parasites. Hilgardia 37:283-335.

Lawson, F.R., R.L. Rabb, F.E. Guthrie and T.G. Bowery. 1961. Studies of an integrated control system for hornworms on tobacco. J. Econ. Entomol. 54:93-97.

Michelbacher, A.E., C. Swanson, and W.W. Middlekauff. 1946. Increase in the population of *Lecanium pruinosum* on English walnuts following applications of DDT sprays. J. Econ. Entomol. 39:812-13.

Olson, W.H. 1974. Dusky-veined walnut aphid studies. Cal. Agr. (July) 1974:18-19.

Rabb, R.L. 1971. Naturally-occurring biological control in the eastern United States, with particular reference to tobacco insects, p. 294-311. *In* C.B. Huffaker [ed.], Biological control. Plenum Press, New York.

Schlinger, E.I., and E.I. Dietrick. 1960. Biological control of insect pests aided by strip-farming alfalfa in experimental program. Cal. Agr. 14:8, 9, 15.

Sluss, R.L. 1967. Population dynamics of the walnut aphid, *Chromaphis juglandicola* (Kalt.), in northern California. Ecology 48:41-58.

Southwood, T.R.E., and M.J. Way. 1970. Ecological background to pest management, p. 6-28. *In* R.L. Rabb and F.E. Guthrie [eds.], Concepts of pest management. N. Carolina State Univ., Raleigh.

Stern, V.M., R. van den Bosch, and T.F. Leigh. 1964. Strip cutting of alfalfa for lygus bug control. Cal. Agr. 18:5-6.

Utida, S. 1957. Population fluctuation, an experimental and theoretical approach. Cold Spring Harbor Symp. Quant. Biol. 22:139-51.

Van den Bosch, R., B.D. Frazer, C.S. Davis, P.S. Messenger, and R. Hom. 1970. *Trioxys pallidus*—an effective new walnut aphid parasite from Iran. Cal. Agr. 24:8-10.

Van den Bosch, R., and P.S. Messenger. 1973. Biological control. Intext Educational Publ., New York.

White, E.G., and C.B. Huffaker. 1969*a*. Regulatory processes and population cyclicity in laboratory populations of *Anagasta kühniella* (Zeller)(Lepidoptera: Phycitidae). I. Competition for food and predation. Res. Pop. Ecol. 11: 57-83.

———. 1969*b*. Regulatory processes and population cyclicity in laboratory populations of *Anagasta kühniella* (Zeller) (Lepidoptera: Phycitidae). II. Parasitism, predation, competition and protective cover. Res. Pop. Ecol. 11: 150-85.

PART V

Insect-Pest Management

PEST MANAGEMENT: ORGANIZATION AND RESOURCES FOR IMPLEMENTATION

Waldemar Klassen

The greatness of John Henry Comstock and Anna B. Comstock continues to fascinate, stimulate, and challenge contemporary entomologists. Because of the pioneering efforts of the Comstocks, Cornell University became a mecca for aspiring entomologists. No one can peruse Comstock's *Introduction to Entomology* without recognizing that Comstock was simultaneously motivated by the wonder of the insect world and by the depredations caused by a small fraction of their number.

In Comstock's time various means for dealing with insects and other classes of pests were being discovered, but resources were meager. Today the federal government spends more than $100 million per year on research and development pertinent to all classes of pests and their management (table 1). The chemical industry invests roughly $70 million per year on pesticide research and development (Ernst and Ernst 1970). State governments also appropriate substantial sums for research on pest management. This research effort is generating a considerable amount of new information, but the rate at which this new knowledge and technology is reduced to practice is extremely slow.

The remedy for this inefficient use of accumulated knowledge, partially developed technology, and research capability was identified and described by Prof. E.H. Smith (1972) as

> a need for a more effective continuum of effort from basic research to actual practice. The links between the various levels of research going from basic to applied are weak. We can hardly afford research on plant resistance, for instance that has as its goal the establishment of differential susceptibilities without regard to the testing of these findings in practice. This in turn requires joint effort by seedsmen, producers, marketers, consumers. The components that form the full spectrum of this effort are frequently not assembled and directed to the goal.

TABLE 1

Federal expenditures in fiscal year 1973 for research and development on pest management

	Pest-Management Technology and Information ($1,000)	Environmental and Health Aspects ($1,000)	Total ($1,000)
U.S. Department of Agriculture			
ARS	59,801	6,483	66,284
APHIS	4,268	0	4,268
CSRS	10,324	1,426	11,750
ERS	340	0	340
FS	12,530	617	13,147
National Science Foundation	3,088	0	3,088
Department of Health, Education and Welfare	1,783	5,984	7,767
Department of the Interior	1,912	3,560	5,472
Department of Defense	764	680	1,444
Environmental Protection Agency	2,378	7,140	9,518
Total	97,188	25,890	123,078

The results are progress reports, satisfying to the individual but without impact on practice. It is an easy way out to work in compartments of basic and applied research and extension without ensuring that bridges are built between them. A team effort that identifies the total spectrum of effort from laboratory to field can overcome this "falling between the categories." Just as interdisciplinary research is required so is there a need for close integration of the various steps from discovery to utility.

For the decade 1950-1960 experienced scientists estimated that insects, weeds, diseases, and nematodes denied us at least one-third of the potential harvest of our crops (U.S. Department of Agriculture 1965)(table 2). The Council on Environmental Quality (CEQ) concluded that crop losses due to insects and diseases in the United States have increased both absolutely and as a percentage of crop value since the 1940s, but that the opposite is true for losses due to weeds (CEQ 1972). In spite of many advances in practical pest management, overall progress is unsatisfactory.

Clearly, many of the improvements in our technology for managing pests have been offset by changes in production practices which may favor pests, such as (1) shift to virtual monoculture of important crops, (2) increased use of irrigation, (3) widespread use of inorganic fertilizers, (4) longer production seasons, (5) widespread use of genetically uniform varieties and (6) reduced tillage. In addition, a number of foreign pests have penetrated our quarantines and have occupied extensive areas.

Furthermore, scientists appear to be in general agreement that our attempts at controlling pests cause enormous impacts on nontarget species. The scanty literature on this topic has been reviewed by Pimentel (1971, 1972). No estimates have been published of the magnitude of these adverse environmental impacts, and no one has even suggested appropriate methods and units of their measurement.

Even though we lack precise quantitative data cherished by budget analysts in evaluating costs and benefits, hard experience has shown that in the long run agricultural productivity cannot be maintained if ecologically unsound approaches are used in dealing with pests. Inappropriate pesticide use has led to (1) the emergence of insect strains highly resistant to insecticides; (2) excessive depletion of predators and parasites and resultant outbreaks of major insect-pest populations and the elevation of minor pests to major-pest status; (3) ecological shifts in weed populations, replacing weeds susceptible to herbicides with species more difficult to control; (4) pesticide residues in commodities, sometimes impeding commerce; and (5) residues in the environment, undermining public confidence in the safety of pest-management practices.

The public is aware of mechanisms such as biomagnification and bioaccumulation whereby certain chemical pesticides

TABLE 2

Estimated average annual losses of potential production of selected plant commodities

Commodity	% Loss of Potential Production by				Total Loss from Pests	
	Diseases	Insects	Weeds	Nematodes	Percent	Quantity[a] 1000 Units
Corn	12	12	10	3 (5)	37	3,270,000 bu
Cotton	12	19	8	2 (5)	41	9,200 bales
Peanuts	28	3	15	3 (10)	34	1,680,000 lb
Grain sorghum	9	9	13	? (6)	31(?)	369,000 bu
Soybeans	14	3	17	2 (10)	36	720,000 bu
Sugarbeets	16	12	8	4 (10)	40	6,940 tons
Alfalfa[a]	20	15	6	3 (5)	44	61,000 tons
Forage seed[b]	15	18	12	5	50	45,000 lb
Apples	8	13	3	? (10)	24(?)	1,830,000 lb
Grapes	27	?	15	?	42(?)	1,820 tons
Oranges	12	6	5	4 (15)	27	70,600 boxes
Lima beans	10	13	8	5	36	51.2 tons
Snap beans	20	12	9	5 (20)	46	658 tons
Potatoes	19	14	3	4 (10)	40	196,000 cwt
Tomatoes (process)	22	7	7	8 (15)	44	4,560,000 tons
Chrysanthemums	11	5	?	?	16(?)	18,400 plants
Shade and ornamental trees	1	11	?	?	12(?)	912 trees

NOTE: Estimates are based on USDA (1965) data from the period 1951-60. Figures in parentheses are taken from the Society of Nematologists (1971).

[a] Based on 1972 production (USDA 1973).

[b] Estimated average losses by National Program Staff, Agricultural Research Service.

applied on farm fields are able to threaten fish and bird populations even in habitats far from treated farmland. Similarly, we have learned that alternative methods of pest suppression can be suboptimal. To facilitate production and marketing, all of our major crops have been placed on narrow genetic bases, predisposing them to potential catastrophic events such as the southern corn-blight epidemic in 1970. The declining thoroughness of animal-husbandry practices appears to have substantially increased the burden on the sterile-male method for managing the screwworm.

These and many other similar experiences led to the realization that systematic ways must be found to utilize optimally all available methods of pest suppression in combinations or sequences appropriate to local situations. Thus pest management is a systems approach to reducing net economic losses from pests and to reducing the impact of pest-suppression measures on nontarget organisms.

A greatly simplified hypothetical scheme of what we hope to achieve through improvements in a practical pest-management system is shown in figure 1. A rough approxima-

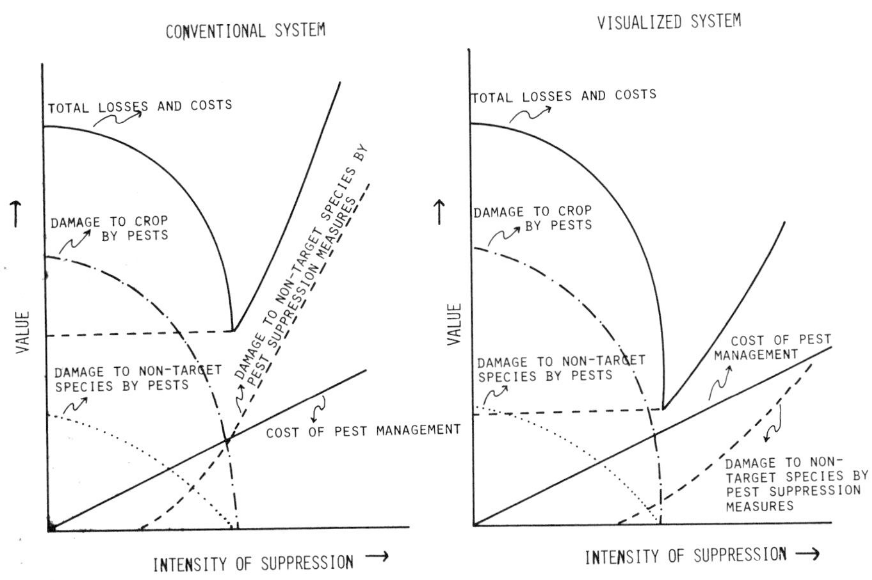

Fig. 1. Hypothetical scheme of costs and losses in the use of conventional and visualized pest-management systems. The benefit of the visualized system is given by a - b.

tion of the total burden to society of a pest is obtained by adding to the cost of pest management the values of damage to the crop by the pest, damage to the environment by the pest, and damage to nontarget species by pest-suppression measures. The benefit from the visualized system is the difference between the sum with conventional pest management and the sum with the visualized system.

In recent years the U.S. Department of Agriculture has been deeply involved in two special pest-management thrusts. The Integrated Pest Management Thrust, initiated in fiscal year 1972, consists of the Huffaker Research Project; the Pilot Research Program; a review of all federal pest-control programs; a program of curriculum development, training, and certification for crop-protection specialists; and an implementation program. The Combined Forest Pest Research and Development Programs, initiated in fiscal year 1975, include the Gypsy Moth Program, the Southern Pine Beetle Program, and the Tussock Moth Program. These Combined Forest Pest R&D Programs were initiated by the Office of the Secretary, USDA. In this paper I limit my discussion to federal contributions to these programs. Nevertheless, I recognize the large input by states and the private sector in developing concepts and individual projects and in providing resources for these programs. In my opinion, these thrusts primarily grew out of the responses of the USDA to six situations.

The first situation began to develop in 1966 when the pink bollworm extended its geographical range to include Arizona and southern California. This pest was suppressed by as many as twenty-five annual pesticide applications, many of which included DDT. Resultant DDT residues in forage and feeds quickly created DDT levels in milk and butterfat in excess of federal tolerances. To counter this problem, the Arizona Pesticide Control Board established a moratorium on DDT usage on agricultural crops in January 1969. Subsequently, Cooperative Extension took the lead in organizing a pest-management program. The successes and difficulties encountered in this pest-management program had a guiding influence on the larger effort supported by the USDA in 1972. In particular, the desirability of avoiding misunderstanding by the pesticide industry was noted.

The second situation instrumental in influencing policy was a series of organophosphate-insecticide poisonings in North Carolina. The Cooperative Extension responded to these fatalities by organizing a pest-management program on tobacco. Also, the North Carolina Department of Agriculture proscribed the sale of mixtures of parathion with other pesticides, thereby reducing sales of this organophosphate by ninety percent, without eliminating availability of parathion for critically important uses (Brazzel, unpublished).

The third situation was the rapid spread of Venezuelan equine encephalomyelitis along the eastern coast of Mexico into Texas. This spread was dramatically halted by a hastily organized insecticide-spray program which decimated mosquito populations over about 13.5 million acres and by a horse-vaccination campaign in which about 2.85 million horses were vaccinated (Animal and Plant Health Inspection Service 1973). Vaccine was administered largely by veterinarians in private practice. It was recognized that the twenty-four thousand private veterinarians in the United States are an extremely important resource and that an equivalent force of private specialists is not available to protect crops from pests.

The fourth situation centered on the hearings conducted as adversary proceedings as part of the cancellation procedure for DDT. In attempting to defend essential agricultural uses of DDT, the Department of Agriculture was unable to dispel fears that these uses constituted an unacceptable hazard to nontarget species, including man. The emergency of managing cotton insects with insecticides that are more acutely hazardous to man than DDT seemed to require the adoption of a strongly managed, interagency pest-management program. The Environmental Protection Agency (EPA) also responded to this urgent situation in fiscal year 1974 by providing $760,000 to the Extension Service for Project Safeguard (Committee on Appropriations 1974). Under this project Cooperative Extension and EPA contacted small farmers, custom applicators, pesticide dealers, and medical personnel to explain safe use of pesticides.

The fifth situation which intensified the interest in and need for reliable and effective pest-management programs involved the sudden realization that our major crops are genetically vulnerable to catastrophic attack by insects and diseases. In 1970 an epidemic of southern corn leaf blight swept rapidly over the nation's corn crop. The yield of corn dropped roughly fifty percent in some southern states and fifteen percent nationwide. Corn is vulnerable to a mutant pest because of its narrow genetic base. Indeed, seventy percent of the seed corn used in 1970 was developed from five inbred lines. Resistance to leaf blight is determined by genetic factors contained in the cytoplasm, and most of the commercial corn had the Texas male-sterile cytoplasm. A committee of the Agricultural Board, National Academy of Sciences (NAS), reported that all of the major crops grown in large monocultures have a very restricted genetic base and are therefore vulnerable to epidemics of diseases or insects (NAS 1972). An ad-hoc subcommittee of the Agricultural Research Policy Advisory Committee (1973) issued a series of recommended actions and policies for minimizing the genetic vulnerability of major crops, including a substantial strengthening of programs of pest surveillance and management.

The sixth situation was created by the current outbreaks of the gypsy moth, Douglas fir tussock moth, and southern pine beetle, and to some extent this problem is related to the lack of adequate substitutes for DDT.

Examination of these situations led to the realization that a high scientific and technological content is inherent in all sound pest-management programs, unlike many public programs. Furthermore, the multifarious local ecological conditions require considerable adaptation of each project. Therefore, a comprehensive pest-management program cannot succeed without many highly qualified people and without both multidisciplinary and multiagency components. The research, teaching, and extension capabilities of our universities and the enthusiastic participation of industry and growers are invaluable in the conception and execution of effective pest-management programs.

It is axiomatic that decision making in public programs should be logical and coherent and involve the general public to the maximum practical extent. Further, increasing emphasis is placed on the quantitative definition of expected accomplishments from public programs. For example, budget analysts and auditors feel that economic and environmental impacts of pest-management projects should be estimated before such projects are undertaken. Once undertaken, programs and individual projects are to be tracked closely to ascertain whether expected accomplishments are being realized on schedule and to assure timely and effective course corrections. Strong emphasis is placed on the selection and development of result-oriented executives to manage or guide public programs (Malek 1974).

Comprehensive pest-management programs are organized and guided with some difficulty, not only because their technological content is high, but also because their essential elements are widely dispersed among various federal and state agencies and in the private sector. Further, the expertise and resources pertinent to pest management in each agency are essential for other functions of the agency. Therefore, there is almost no possibility of developing an authority structure in the form of an agency to encompass all pest-management activities.

However, many informal linkages have developed spontaneously to overcome organizational inadequacies and to facilitate work across lines of authority. Occasionally, urgent situations and scarce funds require the design and overlay of management structures across established lines of authority. Such temporary management structures can be extremely effective, as was dramatically demonstrated by the National Aeronautics and Space Administration, which used project management and matrix organization to rocket the United States from an inferior position to world leadership in the use of weather satellites and in the manned exploration of the moon (Chapman 1973). Usually such management overlays are of limited duration. Obviously, the indefinite perpetuation of controls of interagency project managers would undercut the effectiveness of established agencies.

The design of a management overlay for an interagency thrust is not a simple undertaking. The interagency thrust must be structured to respond effectively to dominant forces within each of the involved agencies, as well as to important forces and factors in its external environment. The former need may be satisfied by erecting a program board of agency administrators, and the latter, by an advisory committee representing major involved interest groups. Finally, there is a widely held conviction that, to be successful, organizations must maximize the differentiation of functions of constituent units. Such differentiation between line and staff functions fragments the authority for decision making and assures the existence of conflict. Thus the structure must be capable of resolving conflict at or near the level where it occurs. Failure to do so precludes the indispensible integration of the functions of specialized units (Lorsch and Lawrence 1972). In pest management we begin with ready-made differentiation because research, extension, and regulatory functions are already handled by separate organizational entities.

The organization of Combined Forest Pest R&D Programs reflects current approaches to the management of public programs.

COMBINED FOREST PEST R&D PROGRAMS

In fiscal year 1975, the Department of Agriculture, through the leadership of the assistant secretary for conservation research and education, greatly intensified efforts to develop and implement technology for dealing with the gypsy moth, tussock moth, and southern pine beetle. For more than a decade after World War II, effective control of the gypsy moth and Douglas fir tussock moth had been attained with DDT. However, the environmental problems associated with DDT led to the restriction of its use to emergencies. The tussock moth infests about 2.4 million acres in the Pacific Northwest, and in 1974 DDT was needed on 435 thousand acres to prevent a potential loss of over $70 million.

The southern pine beetle is the most destructive pest of southern pines. Annual tree kill by the beetle amounts to about 400 million cubic feet of wood, including 1.4 billion board feet of sawtimber, much of which is not salvaged.

The range of the gypsy moth has exploded to the west and south since control operations with DDT were terminated in 1957. In 1973, 1.75 million acres were defoliated. There is an urgent need to reduce losses where the gypsy moth is now well established. Furthermore, at the gypsy moth's current rate of spread the option of preventing the moth from spreading beyond the northeastern corridor will exist for a few years only. Thus the need for decisive action was readily recognized.

Prior to the decision to implement these programs, task groups of more than fifteen federal and state research scientists had defined program objectives and had developed coherent and detailed documentation showing in logical order the specific research and development tasks to be accomplished, as well as the needed people and resources. Planning was accomplished by means of the convergence technique, which was originated by the National Cancer Institute (Carresse and Baker 1967) and later adapted to agricultural research (Bayley, unpublished). The recommendations of various groups, such as the National Gypsy Moth Advisory Council, were available to the planners. Also prior to implementation, the USDA explained its plans to administrators of state agencies, the Association of State College and University Forestry Research Organization, and private-sector representatives.

These programs are planned for four to five years. All three programs will implement the available technological improvements for reduction of losses from pests and develop and evaluate new short- and long-term forest- and pest-management systems to suppress or prevent infestations.

The planned organization of these programs is shown in

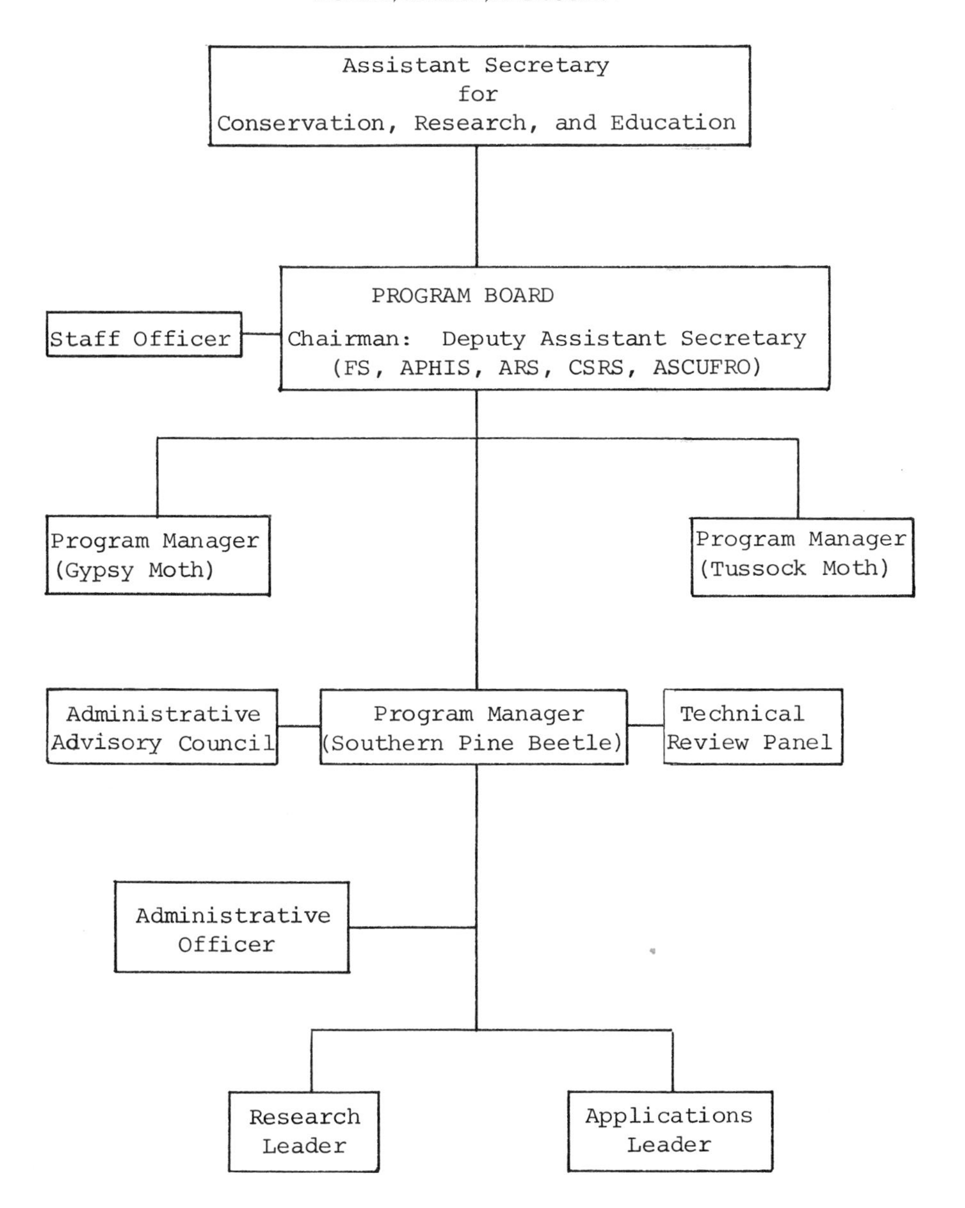

Fig. 2. Organization of combined forest-pest programs, U.S. Department of Agriculture.

figure 2. As experience is gained, changes may be necessary to facilitate accomplishment of objectives. Each program will have a program manager who reports to a program board and specifically to the deputy assistant secretary for conservation research and education, who is chairman of the board. The chairman has operational responsibility for program planning, implementation, review, and control. By means of an interagency agreement, personnel, funds, and facilities will be committed to each program for its duration. Special administrative support will be provided by an appropriate agency. Further, the program manager uses all the instruments of management available to USDA agencies, including cooperative agreements, grants, and contracts with universities, state governments, and private institutions. Funds are allocated in accordance with an annual plan of work and budget approved by the program board.

Key personnel responsible to the program manager include a research leader and an applications leader. The research leader works closely with a technical review panel consisting of USDA and university specialists. The applications leader also works closely with the technical review panel, as well as with knowledgeable federal, state, and private pest-management specialists in designing, conducting, and evaluating field experiments and pilot tests of available and new technology.

An administrative advisory council to advise the program manager is planned for each program. Members of the council are selected from USDA, university, and industry leaders, and representatives are selected from concerned groups and organizations. The chairman of each council reports annually to the board.

The technical review panels screen proposals for research and application and recommend meritorious proposals to the program managers. The panels also monitor progress of funded proposals and recommend appropriate adjustments.

A key element in this organizational construct is the staff officer, who is directly responsible to the deputy assistant secretary. The staff officer reports on the programs to the Office of the Secretary and to the agency administrators. He serves as executive secretary of the program board and arranges annual joint conferences to strengthen coordination, cooperation, and planning.

Clearly, an attempt has been made to integrate the technical and managerial competence of federal and state agencies and universities into a system of participative responsibility. This system provides for central control but

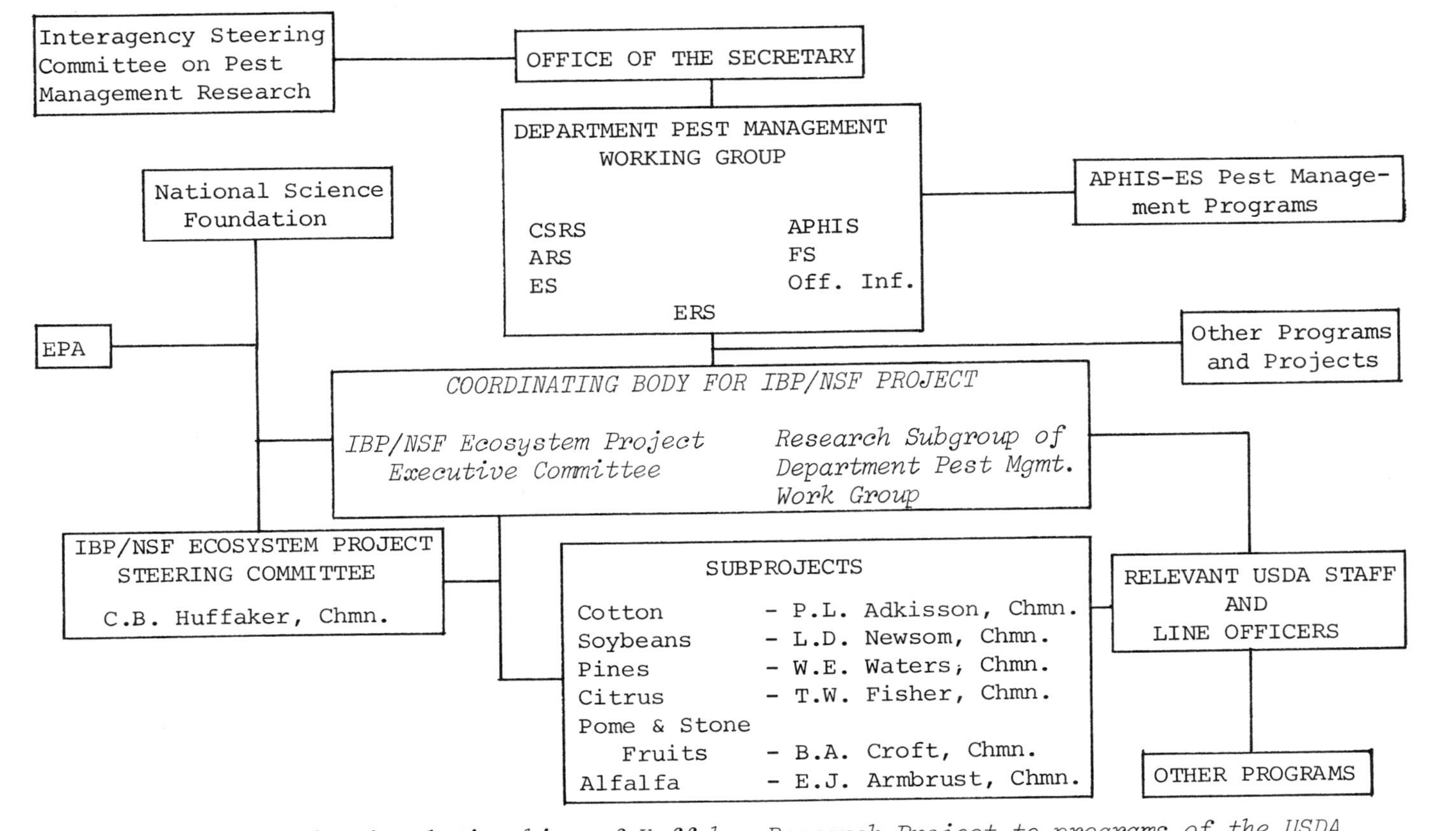

Fig. 3. Organizational relationships of Huffaker Research Project to programs of the USDA and other agencies.

decentralized execution and respects the autonomy of participating agencies. The proposed appropriation for these programs for fiscal year 1975 is $9.3 million, and the projected needs for fiscal years 1975-79 are $46.8 million.

INTEGRATED PEST-MANAGEMENT THRUST OF 1972

In order to implement this pest-management thrust of 1972, the USDA established an Interagency Pest Management Program. Secretary's Memorandum No. 1766 was issued to reaffirm that "the development and use of effective control measures that minimize environmental contamination and related hazards is a high priority objective of the Department in the public interest." Further, the memorandum states that "it is imperative that all available technology be marshalled and applied and that new technology be developed to further progress in achieving this objective." The Interagency Pest Management Program was established to facilitate integrated action by several agencies of the department and to facilitate coordination of the department's efforts with those of the Environmental Protection Agency, the National Science Foundation, and other federal agencies.

Dr. J.S. Robins, former associate administrator, Cooperative State Research Service (CSRS), was designated as manager of the program. The primary mechanism for carrying out this interagency program was the department's Pest Management Working Group. The broad charge given to the working group included (a) the development of cooperation among the agencies regarding the program and especially activities and objectives requiring multiagency effort, and (b) the review of plans, activities, and recommendations of all agencies regarding pest control. This working group was effectively chaired by Dr. J.S. Robins, and key decisions on recommendations of the working group were made by Dr. Ned Bayley, who was then director of science and education in the Office of the Secretary. A research subgroup was appointed to assess research readiness for implementation and to provide technical advice to the working group (see figs. 3 and 4). This interagency program focused primarily on the following five projects:

THE HUFFAKER RESEARCH PROJECT

This research project, entitled "The Principles, Strategies, and Tactics of Pest Population, Regulations and

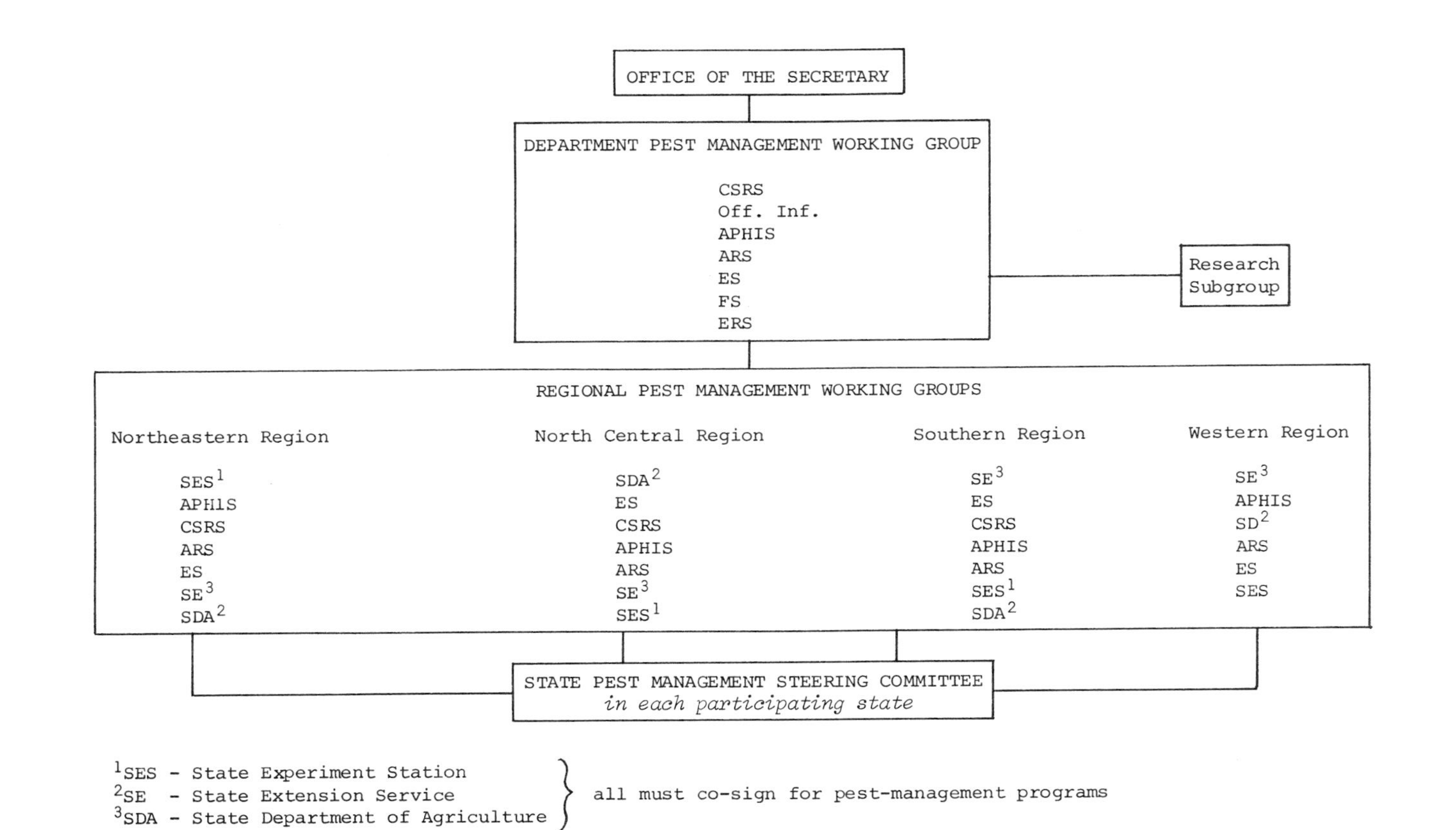

Fig. 4. Organization of USDA pest-management program.

Control in Major Crop Ecosystems," is also known as the Integrated Pest Management Project. Plans for the Huffaker Research Project were developed as a component of the International Biological Program. This program was sponsored by the International Council of Scientific Unions in order to study on a world basis biological productivity in relation to human welfare. The National Science Foundation provided a grant to the University of California to take the leadership in planning and organizing the Huffaker Research Project.

Under the leadership of Dr. Carl B. Huffaker, the project was planned primarily by scientists from nineteen U.S. universities with inputs from scientists in the USDA and in a number of other U.S. and foreign universities. More than two hundred state scientists and about fifty Agricultural Research Service (ARS) and Forest Service scientists participate to some extent in the research. Organization of the project is shown in figure 3. The project is being reviewed for possible funding in fiscal years 1976, 1977, and 1978.

The Huffaker project involves six major crop ecosystems: alfalfa, citrus, cotton, pome and stonefruits, soybeans, and pines. The primary goal of the research is to gain information on a wide range of insect pests that will contribute to their satisfactory management with a minimum of environmental degradation. The goal of the research is to avoid heavy reliance on broad-spectrum insecticides. Depending on the crop and complexity of pests, the investigations include biological-control agents, resistant varieties, cultural measures, attractants, genetic manipulation, and selective insecticides.

The agroecosystems are studied using the methods of systems analysis. Major emphasis is placed on the development of computer-based models for simulation of pest population, crop growth, and their interactions as a basis for selecting, implementing, and evaluating management strategies. Substantial progress has been made.

ARS PILOT RESEARCH PROGRAM

For many years, the need for a vigorous program to develop new methods of pest suppression and detection through large-scale trials has been recognized. Such a program was initiated in fiscal year 1972. Essentially, this program consists of a series of largely in-house projects, most of which are conducted cooperatively with state and industry scientists. A number of projects also are conducted entirely through the agency's extramural program. In the current fiscal year twenty-five projects are under way.

Procedures for developing the Pilot Research Program have been formalized (see fig. 5). Each ARS region has one line

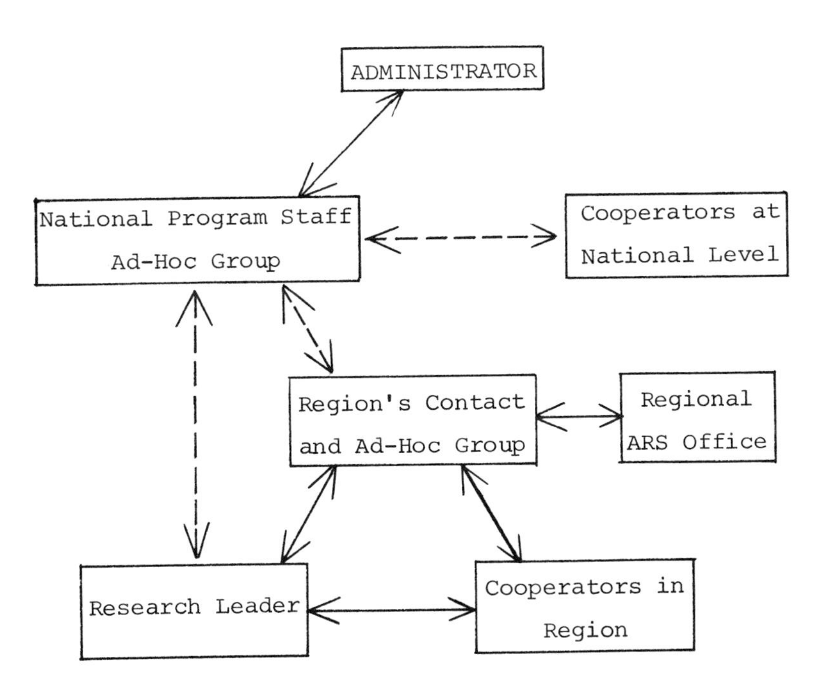

Fig. 5. Interactions in developing program of pilot pest-management research.

official who works with the ARS research leaders and potential cooperators at the state and local level to develop a tentative program of pilot research for the region. Detailed proposals with priority ranking are submitted by each region for review by the national program staff. As part of this review process, the national program staff discusses the projects with representatives of cooperating agencies at the national level. The possibility of submitting the proposed program to the department's Pest Management Working Group for further review is under consideration. Adoption of this procedure would permit other agencies represented on the working group to involve their state counterparts in the review process.

The Pilot Research Program has provided a number of

important accomplishments. It has substantially advanced the use of the gypsy moth pheromone for detection and suppression of low-level populations of this pest. Technology for dealing with the boll weevil has been advanced substantially. A plant-growth regulator has been shown effective in reducing overwintering pink bollworm populations. A sensitive trap has been developed for detection of the Comstock mealybug in California.

REVIEW OF PROPOSED FEDERAL PEST-CONTROL PROGRAMS

The Program Review Panel of the federal Pest Management Working Group systematically reviews the pest-control-program proposals for all federal installations and on all federal real estate. In all cases, special efforts are made to adjust the use of pesticides appropriately and to ensure that the use of nonchemical measures are thoroughly considered.

CURRICULUM DEVELOPMENT, TRAINING, AND CERTIFICATION FOR CROP-PROTECTION SPECIALISTS

The Office of Education of the Department of Health, Education, and Welfare and the Office of the Secretary, USDA, were required "to encourage the development of training and certification programs at appropriate academic institutions in order to provide the large number of crop protection specialists that will be needed as integrated pest management becomes more fully utilized."

The challenge of developing a model Bachelor of Science curriculum was accepted by the Resident Instruction Committee on Organization and Policy (RICOP) with support from the U.S. Office of Education through a grant to BM&M Associates, Inc., Miami, Florida. This committee made a preliminary report recommending a B.S. curriculum and postgraduate programs. Further, RICOP published the proceedings of a workshop in which the concepts for an integrated block of courses in the proposed B.S. curriculum were developed (RICOP 1972).

PEST-MANAGEMENT IMPLEMENTATION PROGRAM

The lead federal agencies in this implementation program are the Animal and Plant Health Inspection Service and the Extension Service. In fiscal years 1972 and 1973, 39 pilot application projects (see table 3) were established in 29 states on 15 commodities (Good, unpublished). Research support is particularly noteworthy for the following projects:

TABLE 3

USDA pest-management pilot application projects

Crop	State	Pests
Corn	Illinois	Insects, weeds
	Indiana	Insects, weeds
	Iowa	Insects, weeds
	Missouri	Insects
	Nebraska	Insects, weeds
	Ohio	Insects, weeds
Cotton	Alabama	Insects
	Arizona	Insects
	Arkansas	Insects
	California	Insects
	Georgia	Insects
	Louisiana	Insects
	Mississippi	Insects
	Missouri	Insects
	New Mexico	Insects
	North Carolina	Insects
	Oklahoma	Insects
	South Carolina	Insects
	Tennessee	Insects
	Texas	Insects
Grain sorghum	Kansas	Insects, weeds
	Nebraska	Insects, weeds
	Oklahoma	Insects
	Texas	Insects, weeds
Peanuts	Oklahoma	Diseases, nematodes, insects, weeds
	Texas	Insects
Tobacco	North Carolina	Insects
Alfalfa	Indiana	Insects
Alfalfa seed	Washington	Insects
Peppers, potatoes	Delaware	Insects, diseases
Sweetcorn, beans	Maryland	Insects
Sweetcorn, lettuce	New Jersey	Insects
Potatoes	Idaho	Insects, diseases
Apples	Michigan	Insects, diseases
	New York	Insects, diseases
	Pennsylvania	Insects, diseases
Apples, pears, peaches	Washington	Insects, diseases
Pears	California	Insects, diseases
Citrus	Florida	Insects, diseases, nematodes, weeds

alfalfa in Indiana; apples in Washington, Michigan, New York, and Pennsylvania; pears in California; citrus in Florida; grain sorghum in Nebraska; peanuts in Oklahoma; peanuts and grain sorghum in Texas; potatoes in Idaho; corn and cotton in various states; and vegetables in Maryland.

The President's fiscal-year-1975 budget (see Committee on Appropriations 1974) would provide increased funds to the Extension Service for the following:

1. to insure continued professional extension support and for some expansion of existing pilot projects into new areas;
2. to initiate new comprehensive pilot application projects for managing the entire complex of pests in various cropping systems, including—
   a. soybean, cotton, and corn cropping systems in the mid-South;
   b. corn, soybean, and small-grain cropping systems in the north-central region;
   c. tobacco, corn, and cotton cropping systems in the mid-Atlantic or southern region;
   d. peanuts, corn, cotton, and small-grain cropping systems in the South or West;
   e. vegetables, alfalfa, and small-grain cropping systems in the Northeast or West;
   f. potato, corn, and small-grain cropping systems in the Northeast or West;
   g. sugar-beet, small-grain, and alfalfa cropping systems in the western or mountain region;
   h. peach-replant and short-life complex in the South or West.

After a three-year period of testing and adapting technology, demonstration projects would be expanded and evaluated and, if accepted, the growers would begin to finance the projects. Extension Service funding might continue for periods of three to six years.

When this program was launched in 1972, four regional pest-management working groups were organized. These regional working groups assisted the USDA in defining requirements for application projects and pilot application projects, in reviewing project proposals, and in ranking them by priority. Further, each participating state organized a state pest-management steering committee to plan, execute, and evaluate the pest-management program or to assist a local steering committee to do so. Composition on these working groups and committees is shown in figure 4.

Under Dr. Robins's leadership, the working group

identified and developed the following requirements for effective programs, program elements, and the functions and responsibilities in application and pilot application programs:

Requirements for Effective Pest-Management-Implementation Projects

1. Adequate technology must exist to assure reasonable success of integrated approaches for managing individual pests or groups of pests. Concurrent research effort to refine the technology and to develop additional technology may increase the probability of success. A written assessment of research readiness should be available prior to initiating any action project.
2. An effective system for transferring available and imminent technology to the ultimate user must exist or be devised. Roles and responsibilities of concerned federal and state agencies must be ordered into an effective system for planning, development, implementation, review, and adjustment. As a minimum, the system has to have the capability of assembling and interpreting current research data and field experience and of reducing that information into a functional project for application at local levels. Additionally, it is hoped that the system will assure (a) that research scientists will challenge extension, regulatory, and industry experts to reduce new technology to practice; (b) that those engaged in implementing new technology will effectively challenge research scientists to conduct operations research in support of implementation efforts and to develop additional technology for possible implementation; (c) that continuous feedback will exist between research and practice so that improvements in technology will be progressively developed through an iterative process of research results advancing to implementation and practical experience triggering further research; and (d) that data and experience obtained will be adequate to validate models of pest suppression and of environmental impacts.
3. Interest, understanding, confidence, and support of the primary users for practical and economical pest-management systems must develop rapidly.
4. Eventually, the general public must develop interest, understanding, confidence, and support for costworthy and ecologically acceptable programs. Support of the general public can be realized only through sensitive measurement

and evaluation of trends of pesticide residues and of impacts of pest-management practices on nontarget species.

Project Elements of Pest-Management-Implementation Program

Three major project elements were identified in a total pest-management program:

The Research Element. An obvious prerequisite for an implementation project is that sufficient knowledge and technology be available to permit substantial modification of current grower pest-management practices. Moreover, in the long run, improvements in grower pest-management practices are dependent on the generation of in-depth ecological and biological information on pests within the changing crop setting so that various new suppression measures may be integrated with natural control factors. Research must point the way, not only to more effective reduction of losses from pests, but also to the amelioration of the impact of pest-management practices on nontarget species.

The Pilot-Application-Project Element. Pilot application projects have two purposes: (1) to refine, test and evaluate available technology in practical field systems; and (2) to provide a practical demonstration that may enhance the rate of adoption of the technology. Pilot application projects require cooperative inputs of federal and state research, educational, and action agencies throughout the planning and development, implementation, evaluation, and adjustment processes. However, the lead responsibilities for these processes lie with educational and regulatory agencies.

The Application-Project Element. A great deal of the information and technology generated through research and experience can be implemented without further refinement in pilot application projects. On the other hand, technology refined in a pilot application project may receive an additional demonstration in an application project. Again, educational and action agencies have the lead responsibility in identifying opportunities, planning and development, implementation, evaluation, and adjustment.

Functions and Responsibilities in Application and Pilot Application Projects

Project Planning and Development. For pilot application projects involving federal initiative, the lead responsibility for planning and project support has resided jointly with the Animal and Plant Health Inspection Service and the

Extension Service with technical support from, and in consultation with, the Agricultural Research Service, Forest Service, Cooperative State Research Service, Economic Research Service, and the National Association of State Departments of Agriculture. As specific needs and opportunities were identified, state and local counterparts and industry inputs have been solicited.

Generally, the cooperative extension services have the major responsibility in counsel with the Extension Service for planning and development of application projects. The Animal and Plant Health Inspection Service, state departments of agriculture, and research agencies should be involved from the outset.

Project Implementation. Successful project implementation uniformly involves the organization of users, provision of educational and information services, technical assistance, monitoring and data acquisitions, and a data-organization management and output system. Administration of funds, training of supervisors and scouts, and other inputs are also involved.

For pilot activities with federal input, implementation is considered the responsibility of the state cooperative extension in cooperation with the state's research and regulatory agencies. Funding for implementation is provided by Extension (USDA) and APHIS. Funding for research components is the responsibility of ARS, CSRS, or the state agricultural experiment station. Recruitment, training, and supervision of scouts and other professional and non-professional staff was considered a responsibility of the cooperative extension service. National data management, storage, and manipulation are the responsibility of APHIS, since this activity relates to quarantine and cooperative pest-suppression programs, the cooperative economic-pest survey, and to the APHIS responsibilities under the International Plant Protection Convention. Research support was developed either through linkages with ongoing research programs or through special allocation of resources by CSRS, ARS, and the state agricultural experiment stations.

Leadership for implementation of application projects was also considered most appropriate for the Cooperative Extension, and all other functions and responsibilities are handled in the same manner as those of the pilot application projects.

Project Review and Adjustment. Continuous review of projects implementing new technology is necessary to ensure that not only the economic goals but also the environmental goals are being met. As pointed out by Byerly (unpublished), "where the Federal dollar goes, the interest of all the people should be considered, not only the interest of the immediately affected interest group."

The working group assumed that project and environmental monitoring and associated data collection and project adjustment would be handled by Cooperative Extension. (However, in some instances Cooperative Extension was unable to collect environmental-impact data, and APHIS assumed this responsibility.) Data-bank output would be provided by APHIS. Review and interpretation of data would be made jointly by federal and state extension, regulatory, and research personnel, often with the Economic Research Service assuming the leadership role. The conclusions and information from the review and adjustment process would then be injected into the planning process by those same agency elements that had handled the initial planning of the projects.

STRENGTHENING THE INTERAGENCY PEST MANAGEMENT PROGRAM

Needless to say, this Interagency Pest Management Program from the outset was beset with concern and controversy. Some have feared that somehow the pest-management thrust was a Trojan Horse for eliminating most pesticides, for field representatives of chemical companies had limited involvement. Others have felt that pest management was simply a new term invented to elicit funds from granting agencies for the support of traditional research and implementation projects. Many have charged that the pest-management thrust was merely an insect-management thrust and that the effort was dominated throughout by entomologists.

Some have pointed out that appropriate mechanisms were not devised for interest groups to register concern and to provide advice. Some state regulatory personnel expressed concern that often the state departments of agriculture are not involved in early planning stages and that regulatory aspects of programs receive inadequate emphasis and support. Some extension personnel have expressed concern that funds to Cooperative Extension are supplied by both the Animal and Plant Health Inspection Service and Extension Service and that the great differences in accounting and certification procedures required by these services are troublesome and require two lines of communication.

Many extension and regulatory personnel feel that the input from research agencies into these projects is inadequate. Even though the reduction of environmental hazard from pest-management practices is an objective of equal weight with that of reducing losses from pests, in practice the environmental objective often has been neglected (Byerly, unpublished).

These problems and others were reviewed in depth for the USDA by Dr. T.C. Byerly (unpublished). Byerly suggested a number of adjustments (see fig. 6) in the program which are currently under consideration. Byerly feels strongly that each participating USDA agency should manage those aspects of programs and projects funded by it. Therefore, the Interagency Pest Management Program should not have a manager, but a coordinator. Coordination at the federal level would be accomplished by a department Pest Management Committee chaired by a staff member of the assistant secretary for conservation research and education. The coordinator would serve as executive secretary of the committee and as a chairman of a subordinate Integrated Pest Management Working Group.

Each USDA agency would fund, as appropriate, only those cooperative activities conducted by its state counterpart. Thus APHIS would no longer fund extension activities. Administrators of the research agencies would be included as signatories of the interagency agreement. Further, Byerly recommends that coordination in the state should be the responsibility of the dean of agriculture in the land-grant university. Finally, Byerly recommends the establishment of a Statutory Pest Management Advisory Committee.

Byerly's specific recommendations reflect the long-established authorities, areas of responsibility, and areas of expertise of regulatory and extension agencies. Byerly recommends that the Extension Service should continue and expand support of the following functions of Cooperative Extension: providing state and area specialists; providing information and training for growers, commercial applicators, consultants, and pesticide-industry employees; recruiting and training scout supervisors and scouts; providing supervision and administrative guidance for scouting activities; planning and conducting application and pilot application projects; collecting and disseminating data for current grower decisions; and providing educational support to facilitate compliance with necessary regulatory and related activities.

Correspondingly, Byerly recommends that the Animal and Plant Health Inspection Service should continue to expand support for the following activities as appropriate: quarantine, condemnation, indemnification, prosecution; prohibition of the planting of pest-susceptible crop

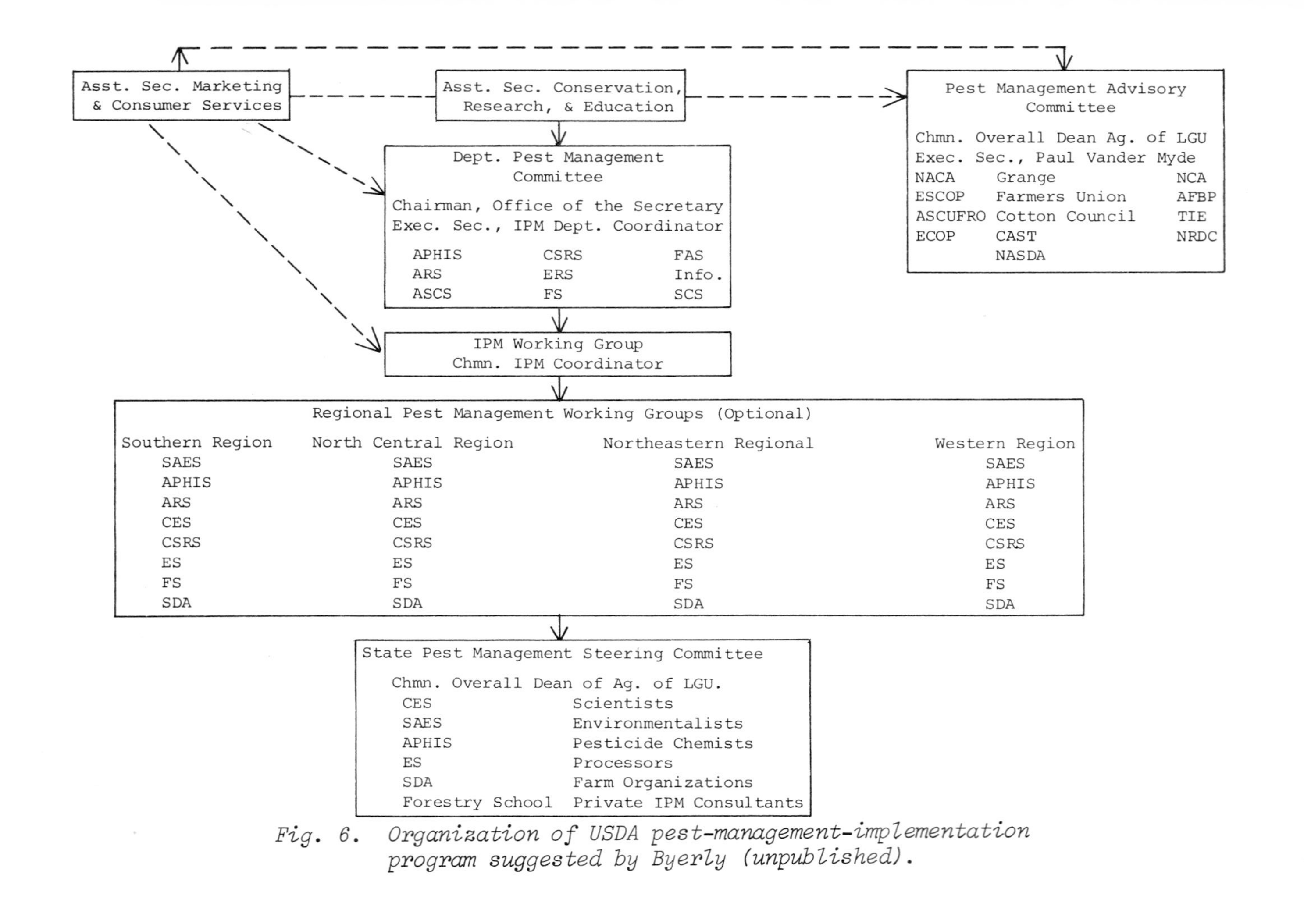

Fig. 6. Organization of USDA pest-management-implementation program suggested by Byerly (unpublished).

varieties; certification of planting stocks for pest freedom; surveillance of pesticide use and application; propagation and distribution of biological-control agents; use of light and pheromone traps for pest detection and suppression; data collection and analysis to validate and implement computer-based models to measure the economic and environmental impact of the project, and to improve methods; providing information on pest populations and identifying problems which require research.

As an alternative, Byerly recognized that assignment of responsibility for USDA-agency leadership could be made on the basis of the degree of regulatory involvement needed to assure effectiveness of the program. For example, those programs which may be effective on a farm-by-farm basis would be assigned to Extension Service, while those which require mandatory area-wide pest-population management would be assigned to APHIS. However, such an arrangement would be likely to engender innumerable disruptive disputes concerning the appropriate strategy.

The input of research agencies into the implementation program have not been given special consideration. Unless the needs of research agencies are recognized in the budget-development process, resources will not be adequate to provide a research component for most pilot application programs. Under the current circumstances, the effectiveness of research agencies cannot be further enhanced in (1) providing research-readiness analyses, (2) anticipating problems before a critical need for action occurs, and (3) initiating research to correct inadequacies prior to implementation of a pilot application project.

A critical weakness in most projects is that they make little use of systems analysis and models needed for proper execution, economic evaluation, and evaluation of impact on nontarget species. Models need to be researched within the context of application programs to integrate measures taken against all classes of pests into the crop-production system. Byerly (unpublished) noted that, unless this is done, "we will continue to have a welter of competing discipline-pest-crop oriented projects just as we have always had."

Finally, it seems to me that the Environmental Protection Agency should be invited to participate as a full partner and on a day-to-day basis in these pest-management thrusts. There may be opportunities to involve the EPA's soil-monitoring program in the appraisal of environmental impact in these programs. Further, EPA involvement could lead to useful discussions of possible mechanisms which could permit uses of pesticides which deviate from the manufacturer's label without jeopardy to the manufacturer. EPA has a responsibility to develop alternate methods of pest management in cooperation with other agencies and to provide educational support to users, as in the case of Project Safeguard.

I do not know to what extent the department will be able to accept Dr. Byerly's recommendations. Certainly, the interagency program can continue to function without a manager or coordinator, as it did in fiscal year 1974. In that case, representatives of extension, research, and regulatory agencies will meet informally several times per year and apprise each other of their respective programs, plans, and needs. However, consequential joint planning between extension, regulatory, and research agencies probably will be restricted to emergency situations.

In the absence of commitments formalized in interagency agreements, it is likely that special research funding through the department's extramural programs to underpin pilot application projects will be very meager. Currently, severe fiscal difficulties experienced by research agencies would appear to preclude the funding of research components even in most pilot application projects.

Extension (USDA) is not a granting agency in the usual sense and is not staffed to handle more than about $2 million per year in nonformula funds for pilot application projects. According to Good,[1] the Extension Committee on Operating Policy has tentatively concluded that in the future additional pest-management application programs could be responsibly managed if funded by formula under the Smith-Lever Act at about $4 million per year. However, such funds have not been requested. APHIS has new responsibilities under the International Plant Protection Convention (ratified by the U.S. Senate and signed by the president in 1972) to improve its pest surveillance and prediction capabilities. Thus in due course APHIS probably will improve its data-collection handling and pest-prediction capabilities.

The task of marshalling our resources to strengthen the continuum from basic research to practice would be facilitated if agreement could be reached that pest losses and adverse impacts should be progressively reduced nationwide an average of, say, ten percent per year. Such a quantitatively defined goal would challenge the nation to mobilize its capabilities. Under emergency conditions in

[1]Personal communication.

1972 and again in 1975 we demonstrated that we can marshall effective research, development, and implementation programs. However, whether we can achieve long-sustained efforts would seem to depend on the development of widespread support and adherence to the concept of closely coordinated multiagency state-federal pest-management programs. Thus, to remain consistent with the tradition of the Comstocks, we will have to set a high standard and get on with the job.

ACKNOWLEDGMENTS

I appreciate the helpful comments on this manuscript made by the following individuals of this department: J.R. Brazzel, J. Good, H.O. Graumann, D.J. Ward, W.B. Ennis, K.R. Shea.

REFERENCES

Agricultural Research Policy Advisory Committee. 1973. Recommended actions and policies for minimizing the genetic vulnerability of our major crops. U.S. Dep. Agr. and Nat. Ass. State Univ. Land Grant Coll.

Animal and Plant Health Inspection Service. 1973. The origin and spread of Venezuelan equine encephalomyelitis. APHIS 91-10. Govt. Print. Office, Washington.

Bayley, N.D. Planning and program structure in agricultural research. Unpublished ms.

———. Evaluating alternative sets of activities with ACTAR planning. Unpublished ms.

Brazzel, J.R. The role of regulatory agencies in pest management. Presented at the Entomology-Plant Pathology Workshop, New Orleans, 1972. Unpublished ms.

Byerly, T.C. Evaluation of Pest Management Project, USDA. Unpublished ms.

Caresse, L.M., and C.G. Baker. 1967. The convergence technique: a method for the planning and programming of research efforts. Manage. Sci. 13:420-38.

Chapman, R.L. 1973. Project management in NASA: the system and the men. Nat. Aeronaut. Space Admin. Govt. Print. Office, Washington.

Committee on Appropriations. 1974. Agriculture-environmental and consumer protection appropriations for 1975, part 3.

Govt. Print. Office, Washington.
Council on Environmental Quality. 1972. Integrated pest management. Govt. Print. Office, Washington.
Ernst and Ernst. 1971. 1970 industry profile study. Nat. Agr. Chem. Ass., Washington.
Good, J.M. Pilot programs for integrated pest management in the United States. Presented at the U.S.-USSR Pest Management Conference, Kiev, 1973. Unpublished ms.
———. The role of extension specialists and private consultants in pest management programs. Presented at the Corn and Grain Sorghum Pest Management Workshop, Lincoln, Nebraska, 1974. Unpublished ms.
Lorsch, J.W., and P.R. Lawrence. 1972. Environmental factors and organizational integration, p. 38-48. *In* J.W. Lorsch and P.R. Lawrence [eds.], Organizational planning: cases and concepts. Irwin, Chicago.
Malek, F.V. 1974. The development of public executives—neglect and reform, p. 230-48. *In* Public administration review, 1974, no. 3.
National Academy of Sciences. 1972. Genetic vulnerability of major crops.
Pimentel, D. 1971. Ecological effects of pesticides on non-target species. Office of Science and Technology. Govt. Print. Office, Washington.
———. 1972. Ecological impact of pesticides. Report 72-2, Cornell University, Ithaca.
Resident Instruction Committee on Organization and Policy. 1972. Systems of pest management and plant protection. Committee on Plant Protection.
Smith, E.H. 1972. Implementing pest control strategies, p. 46-66. *In* National Academy of Sciences, Pest control strategies for the future.
Society of Nematologists. 1971. Estimated crop losses due to plant-parasitic nematodes in the United States. J. Nematol. Spec. Publ. no. 1.
United States Department of Agriculture. 1965. Losses in agriculture. Agricultural Handbook no. 291. Govt. Print. Office, Washington.
———. 1973. Agricultural statistics, 1973. Govt. Print. Office, Washington.

PEST MANAGEMENT:
CONCEPT TO PRACTICE

L. Dale Newsom

Pest management is a term for which no universally acceptable definition has yet been proposed. I define it as an ecologically based strategy for regulating populations of pest species at levels below economic-injury thresholds by use of the most appropriate control tactics available. It may require no more than a single tactic—development of a resistant variety, for example—or it may require a variety of tactics integrated into a highly sophisticated system.

The term *integrated control*, as proposed by Stern et al. (1959), referred to a type of pest management based on achieving maximum effectiveness of natural enemies, supplemented with applications of selective pesticides when necessary. The meaning of the term has since undergone substantial change. One of the most recent definitions is that of Falcon and Smith (1973): "Integrated control is a broad ecological approach to pest control utilizing a variety of control technologies compatibly in a pest management system." In the guidelines they developed for integrated control of cotton pests they listed these "pest control techniques": (1) agronomic practices, (2) biological control by parasites and predators, (3) microbial controls, and (4) chemical controls. The two terms, *pest management* and *integrated control*, are being used to describe essentially the same concept, emphasizing the importance of ecological principles as the basis for pest control.

I propose to discuss pest management within the framework of a statement made by President Lincoln (with a slight modification of my own): "If we could first know [where we have been,] where we are and whither we are tending, we could better judge what to do and how to do it."

An understanding of where we have been in pest management during the last three-quarters of a century is an important guide to the course we should follow now. Time will permit mention of only a few of the events that have led us to our present position. Therefore, at the risk of appearing parochial, I will illustrate pertinent points by drawing examples from two crops: cotton and soybean. They are especially appropriate—cotton, because it accounts for about one-half of all insecticide used in the United States and is rapidly approaching disaster in some areas due to our inability to control some pests; and soybean, because it is a crop in which a key pest has not yet appeared, and thus one that offers unique opportunities for the development of effective, economical, ecologically sound pest-management systems.

COTTON-PEST MANAGEMENT

It is commonly thought that the concept of pest management so widely publicized during the last few years is something relatively new. It has been stated (National Academy of Sciences 1969) that the concept had its beginning about a quarter-century ago in Germany, Nova Scotia, and California. Lest we be led to think that the current concept of pest management is of such recent vintage, let us review some of the early history of cotton-insect control.

Before the boll weevil, *Anthonomus grandis* Boheman, invaded Texas during 1892, there was no key pest of cotton in the United States. The bollworm, *Heliothis zea* (Boddie), and the cotton leafworm, *Alabama argillacea* Hubner, were pests that occasionally caused serious economic damage.

Pierce (1922) outlined a system of cotton-pest management for control of the six species considered major pests and the fourteen thought to be of minor importance. It was reasonably effective. The major components of the system were—

1. winter plowing to destroy pupal cells of overwintering lepidopterous pests;
2. as early as growing conditions would allow, planting rapidly fruiting and early maturing varieties;
3. treating infestations of the cotton leafworm and grasshoppers with powdered arsenate of lead;
4. harvesting the crop as soon as possible and destroying the crop residues by plowing or grazing;
5. reducing populations of hibernating insects by cutting and burning vegetation around fence rows and field roads.

Clearly, this fits our definition of pest management, although it was much less sophisticated than some of the systems now practiced. Two elements of the system were especially effective: early planting of rapidly maturing varieties and early harvest and crop-residue destruction.

Prof. Dwight Isely of the University of Arkansas, whom I

consider the father of pest management in the United States, formulated concepts of pest control and developed an effective system of management of cotton-insect pests soundly based on the principles of applied ecology during the 1920s and 1930s (Isely 1924, 1926, 1934). He was the first to integrate use of an insecticide—calcium arsenate—with the principles of economic thresholds and trap crops into a system of pest management. It was he who first developed a system of using commercial cotton-insect scouts, initiated in Arkansas during the late 1920s.

Professor Isely built his system around an intimate knowledge of the biology and ecology of the key pest—the boll weevil. He recognized the long host-free period as being the most critical in the seasonal history of the pest. Although unaware of the phenomenon of diapause in the boll weevil, he knew that the part of the population becoming adult in September and early October hibernated more successfully than that part maturing earlier or later. Taking advantage of the fact that weevils emerging from overwintering quarters congregate in the earliest fruiting cotton, he planted extra-early maturing, rapidly fruiting varieties as trap crops in areas near favorable hibernation quarters for the weevil. The weevils attracted to the trap crop were destroyed by applications of calcium-arsenate dust, three to four times at five-day intervals. Isely also developed the "spot dusting" technique for taking advantage of the extremely localized distribution of overwintered and first-generation boll weevils.

Using trap crops and spot dusting, it was often possible to limit the amount of acreage treated in a season to five percent. Treatment of such limited areas conserved populations of predators and parasites, had minimal impact on other nontarget organisms, and held environmental pollution to low levels. Isely's system took maximum advantage of the suppressive effects of hot, dry summers and cold winters, climatic conditions to which the boll weevil is highly susceptible.

The system included—

1. use of early fruiting, prolific, rapidly maturing varieties and the appropriate cultural practices required to produce a crop during the shortest possible time;
2. use of trap crops of extra-early maturing varieties to concentrate overwintered boll weevils on a small percentage of the total acreage;
3. spot treatment with calcium-arsenate dust of the

relatively small areas into which overwintered boll weevils tend to congregate, in order to reduce the amount of acreage requiring treatment;

4. a system of scouting at weekly intervals for population assessment in order to employ the principle of economic thresholds when it was necessary to use calcium arsenate to suppress populations;
5. imposition of the hazard of food shortage on the potential overwintering boll weevil population by harvesting the crop and destroying crop residues as early in the season as possible;
6. use of land clearing and controlled burning to reduce the area of favorable hibernation quarters.

This system, evolved a half-century ago, contains the foundations upon which current pest-management systems for cotton-insect pest control rest. Its practice allowed the cotton industry to survive for more than fifty years with little use of insecticides.

In spite of the demonstrated effectiveness of calcium-arsenate dust for control of the boll weevil, most growers were disappointed with its performance and it was never widely accepted. In general, growers that did use calcium arsenate did not use accurate pest-population assessment as the basis for treatment. Often treatments were started too early or too late. When applied to large acreages, calcium arsenate decimated populations of natural enemies, often inducing serious outbreaks of the cotton aphid, *Aphis gossypii* Glover, and *Heliothis zea* (Boddie), and causing more severe damage than that caused by untreated populations of the boll weevil. Because of these problems, most growers never developed confidence in their ability to control cotton pests (except the leafworm) effectively and economically with calcium arsenate.

Prior to the calcium-arsenate era much excellent research was devoted to the biology and ecology of various cotton pests (Hunter and Hinds 1904; Hunter and Pierce 1912; Quaintance and Brues 1905; and Garman and Jewett 1914). After 1920 research of this sort was largely discontinued, and efforts were concentrated on making calcium arsenate a panacea. A high percentage of research during the following quarter-century was devoted to the evaluation of calcium arsenate and various additives to it.

The advent of DDT after World War I began a period of dramatic change in entomology. I well remember a discussion that took place in the basement of Comstock Hall (Cornell University) after publication of Dr. Clay Lyle's presidential address delivered at the 58th Annual Meeting of the American Association of Economic Entomologists. In his address Dr. Lyle stressed the potential of DDT and benzene hexachloride for eradication of a large number of insect pests, saying: "The entomologist has become a wizard in the eyes of the uninitiated—and indeed some of his achievements seem little short of magic"; and, "We have the technical knowledge and equipment to eradicate the house fly, horn fly, cattle grubs, cattle lice and several other insects." The general reaction of the group of graduate students discussing these views with the late Dr. H.H. Schwardt was one of dismay that so prominent an entomologist could believe that the synthetic organic insecticides offered a means of eradicating a large number of pest species.

Nevertheless, many entomologists shared his beliefs and the next quarter-century was characterized by excessive, and almost exclusive, use of the synthetic organic insecticides for insect control. The new insecticides were spectacularly effective in quickly killing a wide spectrum of insect species, unlike calcium arsenate, which killed slowly and with far greater selectivity. The easily observed ability to kill pests quickly and the low cost of these new insecticides promoted their enthusiastic acceptance and widespread, heavy use by growers.

These powerful new chemicals were so effective that the pest-management practices largely responsible for survival of the cotton industry during the period 1892-1945 were no longer thought necessary. The new insecticides were so dependable and economical that cotton breeders began to develop varieties that would continue to fruit during a long period. These new varieties also responded well to high levels of plant nutrients and irrigation water. Heavy rates of fertilizer applications became standard procedure. Such practices, adopted as a means of obtaining higher yields, were eminently successful; yields were more than doubled. Unfortunately, they provided more favorable conditions for the boll weevil, and other species as well, to reproduce at maximum rates for one or more additional generations than had been possible previously. Thus pest problems were substantially intensified.

Results of a large-scale field trial in Wharton County, Texas, in 1948 (Ewing and Parencia 1949) started a disastrous trend in cotton-insect control in the United States. The purpose of the experiment was to test the concept of controlling overwintered boll weevils before they could reproduce. Their objectives also included control of so-called early-season pests such as plant bugs and thrips. They made two applications of insecticide, the first timed to coincide with the beginning of squaring (flower-bud formation) and the second eight to ten days later. Results were excellent. Adequate season-long control of the boll weevil was obtained, and treated areas outyielded the untreated areas by almost fifty percent. Unprecedented publicity was given to this experiment, and growers put the method into practice immediately and widely, even though the early literature showed that it could succeed for boll-weevil control only during rare years when weather conditions during the fall and again the following spring were such that nearly all of the overwintered boll-weevil population could emerge from hibernation quarters and enter cotton fields by or before the time the cotton produced one-quarter-grown squares (flower buds). Emergence from hibernation quarters normally extends over a period of several months. Thus substantial numbers of overwintered weevils entered cotton fields long after insecticide applications, timed to coincide with early fruiting, had lost effectiveness.

This weakness quickly surfaced during subsequent years, and the program evolved into an "automatic" schedule of weekly applications, culminating in what I choose to call the "womb to tomb" program of cotton-insect control—weekly applications of heavy rates of broad-spectrum insecticides from the time cotton emerged to a stand until the crop matured beyond the stage of susceptibility to insect attack. It was generally believed that a panacea had been found. The ultimate in entomological irresponsibility—complete disregard for all principles of applied ecology and common sense—had been reached. A system that is the antithesis of pest management had been accepted and put into practice throughout the Cotton Belt.

This program was popular with growers because it was initially effective, cheap, and exceedingly simple. It was attractive to the pesticide industry because it provided a huge and predictable market for their products. The disastrous results of such a single-faceted system are well known. Briefly, they are—

1. development of resistance to available insecticides by most species of major pests;
2. changes in pest status and species dominance with the change from calcium arsenate to the chlorinated insecticides and from the latter to the phosphates and carbamates;
3. adverse effects on nontarget species;

4. widespread environmental pollution with massive residues of long-persisting insecticides;
5. unacceptably high costs of control;
6. loss of confidence in the entomological profession.

There are now large areas of the Cotton Belt where the continued profitable production of the crop is seriously threatened by induced pests such as the tobacco budworm, *Heliothis virescens* (Fabricius), and the bandedwing whitefly, *Trialeurodes abutilonea* (Haldeman). These pests have developed such high levels of resistance to all available pesticides registered for use on cotton that they must be considered virtually out of control in some areas. At best, control is mediocre and costs of control are so high that they cannot be borne by the crop at current cotton prices, less than fifty cents per pound.

The situation has degenerated beyond the crisis point, as described by Falcon and Smith (1973). On 11 October 1974 Dr. Dan Clower, the Project Leader of Cotton Insects Research in the Department of Entomology, Louisiana State University, announced to a group of county agents from the largest cotton-producing area of the state that the department could no longer recommend any currently available insecticides for control of the tobacco budworm. All registered insecticides, or mixtures of insecticides, have failed to control this pest in large areas of Louisiana during the latter part of the 1974 growing season. The cotton industry in Louisiana appears to be entering the beginning phases of disaster similar to that in the Tampico-Mante area of Northeastern Mexico during 1969 and the Rio Grande Valley of Texas during 1970 (Adkisson 1972).

The tobacco-budworm and bandedwing-whitefly problems are spreading and intensifying in the cotton-growing areas of the mid-South. The severity of the problem presents entomologists with the necessity for developing and implementing a pest-management system that will: (1) control the boll weevil, (2) relax the selective pressures of insecticides on tobacco-budworm and bandedwing-whitefly populations to prevent higher levels of resistance, and (3) allow maximum impact of natural control agents and climatic conditions for suppression of all pest species.

SOYBEAN-PEST MANAGEMENT

The need to control soybean-insect pests is relatively new in the United States. Prior to 1960 a very high percentage of the soybean crop was produced in the midwestern and north-central states. Insect pests of soybean in those areas are few and unimportant. Increasing need for oil and protein has been responsible for an unprecedented expansion in acreage devoted to the production of a major crop in the United States. Acreage planted to soybean increased from 23.7 million acres in 1960 to 56.4 million acres in 1973. Much of this increase, in some cases as much as tenfold, has been in states along the Gulf and South Atlantic coasts, where the pest hazard is extreme. Geographic location and climatic conditions in these states favor long growing seasons and heavy insect infestations. This has proved true for soybean.

The soybean crop is affected by a large complex of insect pests and insect-transmitted virus diseases. All parts of the plant—roots, nodules, stems, leaves, and fruit—are attacked by one or more species of pests. However, no key pests have developed yet. In many respects soybean now is remarkably similar to cotton prior to invasion of the United States by the boll weevil. Like cotton, soybean has a remarkable ability to compensate for substantial amounts of injury by pests without adverse effects on yield or quality.

Thus far little insecticide has been used on the crop. Clearly, it should be possible to avoid making the mistakes in controlling soybean pests that have been so tragically and senselessly made in our attempts to control cotton insects during the last half-century. Nevertheless, the stage is being set for a repetition of these same mistakes with a crop planted to a much larger acreage than was ever planted to cotton. In addition, a high percentage of the acreage planted to soybean is in areas much more sensitive to environmental insult than those planted to cotton.

Because of the huge acreage involved, the insecticide industry is attracted to the soybean as a potential market for huge amounts of chemicals. Because of the high price of soybean, growers will not even tolerate levels of infestations that are far below economic-injury thresholds. Insecticide salesmen are using this argument effectively. The same arguments are being used now as were used in the early days of the "womb to tomb" system of cotton-insect control. Soybean growers are being bombarded with the following sorts of statements and questions: The only good bug is a dead bug. If an insect is feeding on a soybean plant, he is costing you money. Why wait until an infestation reaches the economic threshold? Suppose we were to get a week of rainy weather and could not get into the fields to treat. Why take a chance with insects when soybean is selling for nine dollars per bushel? Saving one bushel per acre will return two dollars for every dollar invested.

CURRENT SYSTEMS FOR MANAGING COTTON AND SOYBEAN PESTS

I have reviewed where we have been, where we are now, and whither we are tending for both crops. We now come to the questions of what to do and how to do it. What to do may be answered immediately and confidently. The strategy is the same for both crops; namely, the development of ecologically based pest-management systems. Deciding what tactics to use and how to implement pest management is more difficult.

An effective pest-management system has been developed for cotton. It has been tested and proved in numerous large-scale field trials across the Cotton Belt, where the boll weevil is the key pest (Brazzel 1961; Lloyd et al. 1966). The key component of the system is destruction of a high percentage of the potential overwintering population of the boll weevil. Only that portion of the population that enters diapause toward the end of the growing season is capable of overwintering. If cotton is treated with appropriate insecticides at ten- to fourteen-day intervals after the crop matures, if a defoliant or dessicant is applied to remove foliage and immature fruit at the proper stage of plant maturity, and if the crop is harvested quickly and the crop residue destroyed immediately by plowing, a very high percentage of the diapausing population is either killed by the insecticide or starved. The total boll-weevil biomass capable of surviving the winter and several months without food is narrowly concentrated in space at this particular season. Thus it is especially vulnerable during the few weeks before it leaves the field for winter quarters.

The system may be made even more effective by use of trap plots, as previously described. Boll weevils congregating in such trap plots may be destroyed by applications of appropriate insecticides. Treatments of such small areas have minimal adverse effects on natural enemies.

Where this system is used effectively, the boll-weevil population is suppressed the subsequent year to the extent that economic-injury thresholds are rarely reached until the appearance of the third generation. This allows maximum utilization of the natural-enemy complex for suppression of populations of *Heliothis* spp. and the bandedwing whitefly. It prevents or delays substantially the need for insecticidal control of these pests. By making applications of insecticides unnecessary until late in the season, prey for the natural-enemy complex consisting of aphids, thrips, spider mites, whiteflies, plant bugs, and eggs and larvae of *Heliothis* spp. can be maintained at populations sufficient to support predators and parasites at levels capable of regulating some pest populations at subeconomic levels.

This relatively simple pest-management system has resulted in a decrease by about one-half in the amount of insecticide required to control cotton pests. It has the further advantage of allowing fewer generations of *Heliothis* spp. and bandedwing whitefly to be exposed to selective pressures of heavy applications of insecticides. It should prevent further deterioration of cotton-insect control until a more sophisticated system of pest management can be developed. At the very worst, it is a vast improvement over the system of total reliance on repetitive applications of broad-spectrum, highly toxic insecticides.

In spite of the amply demonstrated effectiveness of this relatively simple pest-management system and its enthusiastic recommendation to growers by research and extension-service personnel in several states, its acceptance and implementation have been disappointingly slow.

An interesting feature of this system is its remarkable similarity to the system Isely developed a half-century ago in Arkansas. After decades of virtually complete disregard for the concept of pest control based on the principles of applied ecology, cotton entomologists have come full circle.

This system is so effective that it was made the principal component of a large-scale boll-weevil eradication experiment in southern Mississippi. Additional tactics included in this ambitious experiment were: (1) planting a part of the acreage to a variety resistant to the boll weevil, (2) treating the early planted trap crop with a heavy rate of the systemic carbamate aldicarb, (3) trapping overwintered weevils by use of the "pheromone" Grandlure, (4) treating fields having high numbers of overwintered boll weevils with conventional phosphate insecticides in order to reduce populations to levels at which the Grandlure traps operate most efficiently, and (5) use of weevils sterilized by treatment with busulfan to "overflood" the eradication zone and prevent survivors of previous treatments from reproducing.

Soybean insect-pest control is at a critical period of development. The tremendous recent increase in acreage devoted to production of soybean has forced the crop into areas of the South where the pest hazard is high. Fortunately, the potential hazard to soybean posed by insects in the South was recognized soon after the trend toward large increases in acreage became obvious. A regional research project (S-74) was initiated during 1968 to establish priorities for research and to coordinate the research effort. Subsequently, soybean was included as one of the six major crop ecosystems of the NSF/EPA Integrated Pest Management (IPM) Project, often referred to as the "Huffaker Project." The objective of this project is "to develop new, reoriented, expanded, and closely coordinated research efforts seeking practical alternatives to the extensive use of broad-spectrum toxic chemicals for control of certain pest complexes." Nineteen universities plus segments of the Agricultural Research Service and Forest Service, U.S. Department of Agriculture, Environmental Protection Agency, and some elements of private industry joined in this major effort. Six state universities, plus supporting elements from the other agencies, undertook research on the soybean subproject. Many of the personnel of the soybean subproject are also participants in Regional Project S-74. Thus the research on both projects has been a closely coordinated, cooperative effort.

The major objective of the soybean subproject is to develop ecologically based, dynamic pest-management systems that regulate populations of a large complex of insect pests at noneconomic densities, optimize cost-benefit relations, and minimize environmental degradation. Excellent progress has been made, considering the dearth of information available when these coordinated, cooperative efforts were undertaken and the short period of time they have been in operation. Provisional pest-management systems for control of soybean pests have been developed and are being tested in large-scale field trials. Their rapid development has been made possible by close cooperation with, and free flow of information from, the IPM Project.

The systems are relatively simple and vary from region to region, depending on the pest complex involved. The basic components of all are more accurate economic thresholds, improved techniques of population assessment, use of selective insecticides and nonselective insecticides used selectively to make maximum use of native natural enemies, and use of trap crops of early planted, rapidly maturing soybean to control such pests as the bean leaf beetle, *Cerotoma trifurcata* (Forster), and the bean pod mottle virus that it transmits; the Mexican bean beetle, *Epilachna varivestis* Mulsant; and the southern green stink bug, *Nezara viridula* (Linnaeus). The systems use higher economic thresholds for most pest species than formerly were used. This has been made possible by determining the amount of food consumed by one individual of a particular pest species during its development, using more accurate methods for assessing population densities, and having better information on the soybean's ability to compensate for injury by insect pests. Minimum effective dosage rates for specific pests are used, and organochlorine insecticides are eliminated.

The following provisional pest-management system for soybean has been developed for Louisiana: A rapidly maturing variety is planted about ten to fourteen days before the main crop on a relatively small percentage of the total acreage. This trap crop attracts and concentrates a large percentage of overwintering bean leaf beetles (the major vector for bean pod mottle virus) and, during early August, a large percentage of the stink-bug populations from adjacent areas. Both species can be controlled before large populations develop and disperse from the trap-plot areas. Trap plots are treated with 0.125 pound azinphosmethyl per acre for control of bean leaf beetle and 0.25 pound per acre of methyl parathion for control of southern green stink bug. These insecticides, applied at such low dosage rates to a small percentage of the total acreage in a field, have minimal impact on populations of native natural enemies. Even if all the natural enemies are killed in the treated plots, reinvasion from untreated areas is rapid, and populations of beneficial insects are rapidly restored to pretreatment levels.

Economic-injury thresholds for the lepidopterous defoliators including green cloverworm, *Plathypena scabra* (Fabricius); soybean looper, *Pseudoplusia includens* (Walker); and velvetbean caterpillar, *Anticarsia gemmatalis* Hubner, have been raised as information has become available on their damage potential. Natural control agents almost invariably suppress green-cloverworm populations to subeconomic densities. Both the soybean looper and velvetbean caterpillars are annual migrants into Louisiana. The soybean looper reaches economic-injury levels only in those areas where soybean is grown in rotation with cotton. This species is dependent upon a source of carbohydrate to produce the normal complement of eggs. Floral and extrafloral nectaries of cotton provide ample quantities of carbohydrate, and where both crops are grown in close proximity the soybean looper is often a severe pest. In other areas where soybean is grown in a virtual monoculture, or rotated with non-nectar-secreting crops, populations of the soybean looper do not reach economic thresholds and can be ignored. Even when grown with cotton, it is under heavy attack by a large complex of predators, parasites, and microbial pathogens and often does not require insecticidal control. When control is required, low dosage rates of methomyl or *Bacillus thuringiensis* are used.

The velvetbean caterpillar is also an annual immigrant into the state. Although it is attacked by a wide variety of parasites, predators, and microbial pathogens, it frequently overwhelms these natural enemies and requires treatment. It is easily controlled by minimum dosage rates of the conventional pesticides, such as carbaryl, methyl parathion, and the microbial insecticide *B. thuringiensis*.

Other species rarely cause enough damage in Louisiana to require insecticide treatment.

Preliminary testing of this simple, provisional pest-management system in large-scale trials indicates that it can hold insecticide use on soybeans in Louisiana to an average of not more than one application per season to one-fourth, or less, of the total acreage planted to soybean. However, as is the case with cotton, there are already indications that the system will not be accepted and put into practice as rapidly as is desirable. The reasons, similar to those for cotton, will be discussed later.

IMPROVED STRATEGIES FOR MANAGING COTTON AND SOYBEAN PESTS

It is fortunate that it will not long be necessary to depend on the simple, provisional pest-management systems just described. Some exciting new techniques for control of insects have been discovered during the last two decades, and some tried and proven techniques from that long-ago era when pest management was ecologically based have been rediscovered. Some of these are ready to be integrated into existing systems, and others are in various stages of research and development. Among those that are likely to be useful in the near future are: (1) varietal resistance, (2) microbial pathogens, (3) cultural controls, (4) selective insecticides, and (5) pheromones.

Varietal Resistance

Some of the most exciting work on cotton-insect control is that of Knox Walker, G.A. Niles, and J.R. Gannaway of the Texas Agricultural Experiment Station on "fast fruiting" cotton varieties. Their work during 1974 with rapidly fruiting varieties in Frio County, Texas, an area considered "one of the most severe boll weevil areas in Texas" and recently affected by serious bollworm and tobacco-budworm problems, showed the value of rapidly fruiting, prolific, early maturing cotton varieties. They compared the performance of their Tamcot SP-37 with conventional Delta types, using narrow rows, 25 pounds of nitrogen per acre, one irrigation, and two applications of insecticides. Normal practices by growers in the area were wide rows, 100 or more pounds of nitrogen per acre, two irrigations, and six to nine insecticide applications. The Tamcot SP-37 was completely harvested by August 31, the day growers began harvesting the commercial varieties. It outyielded conventional Delta varieties, producing 795 pounds of lint per acre.[1] Use of such varieties, capable of producing and maturing high yields before the boll weevil can produce a second generation, in combination with an effective reproductive-diapause program, imposes tremendous suppressive pressures on boll-weevil populations with minimal use of insecticides.

Other highly promising characters for resistance to cotton pests have been discovered and are in advanced stages of testing. Frego bract, a mutant with straplike bracts instead of the normal large leaflike structures, confers very useful levels of resistance to boll weevil but unfortunately carries with it added susceptibility to the plant bugs *Lygus* spp., *Pseudatomoscelis seriatus* (Reuter), and *Neurocolpus nubilis* (Say). Super okraleaf, a prolific, early maturing variety, can be planted as a trap crop on a small part of the acreage at the same time as the main crop varieties. Using varieties in which the nonpreference character of Frego bract has been combined with that for the red-leaf character has made it possible to delay insecticide applications for boll-weevil control for more than one month under conditions of severe boll-weevil pressure.[2]

Research is progressing toward developing varieties of cotton without nectaries to control plant bugs and some species of lepidopterous pests, as well as the soybean looper in areas where it is dependent on cotton as a source of nectar.[3]

Varietal resistance shows excellent possibilities for control of soybean pests. Varieties in which high levels of resistance to Mexican bean beetle and useful levels of resistance to the bean leaf beetle and soybean looper have been combined are in advanced field-scale trials. At least one is expected to be named and released during 1975.[4]

---

[1]Knox Walker: personal communication.

[2]D.F. Clower and J.E. Jones: personal communication.

[3]Fowden Maxwell: personal communication.

[4]Sam Turnipseed: personal communication.

Microbial Pathogens

Progress in research on microbial pathogens has been slowed during the last few years as a result of the extreme difficulties encountered in registering them for use. It now appears that a well-coordinated effort will be made to speed up the research required to complete registration for some of the nuclear polyhedrosis viruses and granulosis viruses. The *Autographa* virus for control of cotton pests and the NPV for control of soybean looper have proved useful pathogens, apparently capable of perpetuating themselves in crop ecosystems once they have been introduced. More effective formulations of *Bacillus thuringiensis* have been developed and are proving useful, particularly for control of soybean pests.

Of all the natural control agents, the microbial pathogens most nearly resemble the conventional insecticides in ease of manipulation. None other can be brought to bear so quickly on outbreak populations. Thus they could be extremely useful components of pest-management systems.

Cultural Controls

Rediscovery of the importance of cultural control tactics such as early planting, destruction of crop residues, and use of trap crops has provided highly effective components of pest-management systems for soybean and cotton. Trap crops especially may prove to be much more useful than they are currently known to be when the behavioral mechanisms that make them effective are understood.

Selective Insecticides

Large differences in insect-species response to insecticides have been demonstrated. Thus far, no insecticide has been found with enough selectivity to discriminate between pest and beneficial species to any useful degree. Such selectivity remains a much-hoped-for future development.

Although discovery of useful levels of "physiological" selectivity in insecticides appears to be only a remote possibility, "ecological" selectivity has been demonstrated to be a useful tactic. It is effective in pest-management systems that include early planted trap plots or planting a small percentage of the total acreage to a preferred variety, to attract and concentrate overwintered populations in small areas of the total crop where pests can be treated with conventional insecticides, and the remainder to a non-preferred variety.

Pheromones

Isolation, identification, and synthesis of chemicals that affect insect behavior in remarkable ways appear to offer powerful new tools for insect control. Thus far these pheromones have proved to be useful mainly for monitoring populations. In a few cases, they have been used to suppress populations of major pests.

Recently, the "aggregating and sex pheromone" Grandlure has been used as a component of the cotton-insect pest-management system (Tumlinson et al. 1969). It was originally found effective in attracting and trapping low populations of overwintered boll weevils before cotton began fruiting. More recent work (Mitchell and Hardee 1974) indicates that Grandlure-baited traps are effective when used concurrently with insecticides during midseason, when cotton is fruiting heavily.

Such chemicals offer promise for the near future as more effective methods of using them are discovered.

FUTURE PROGRESS

Pest-management systems for cotton and soybean may be unsophisticated, but there is reason for optimism that they can halt and perhaps even reverse the accelerating trend toward disaster in cotton-insect control and can prevent soybean-insect control from following the path taken by cotton. Widespread implementation of existing pest-management programs would put cotton-insect control on a sound basis and establish soybean-insect control so that there would be little or no necessity for use of conventional insecticides. However, substantial obstacles to the successful practice of pest management are developing.

Shortage of Trained Personnel

There are not enough properly trained personnel to staff the pest-management programs effectively. Universities have reacted slowly to the responsibility for training personnel. There has also been widespread lack of understanding of the type of training needed. Our profession lacks the necessary numbers of "general practitioners," well-trained applied entomologists required to operate the pest-management programs already initiated. Unless there is rapid acceleration in training entomologists with enough understanding of agricultural ecosystems to justify their being called pest-management specialists, implementation of pest-management systems will be seriously handicapped.

Growers need pest-control consultants and professional scouting services. They currently are relying far too much upon insecticide salesmen, aerial applicators, and other persons with vested interests in selling insecticides and services.

Adverse Attitudes

Most personnel involved in manufacture, sales, and application of conventional insecticides look upon pest management as a direct threat to their livelihood. They view it as inevitably causing a substantial decrease in the amount of conventional insecticides used. This is a correct assumption. Otherwise, pest management is a failure.

A typical attitude was that expressed last summer by the owner of a large agricultural flying service to one of our extension entomologists. This man, in an angry outburst, said, "Your d—n pest-management program is ruining my business. I can't get pilots who are willing to work for me when there is so little insecticide being used." This, of course, was a fine tribute to the success of the pest-management program, but it also illustrates the extent of the problem. Far too many insecticide salesmen, formulators, distributors, and manufacturers have the same attitude. Unfortunately, these people have intimate, continuing contact, and thus much influence, with growers.

The need has long been recognized, and has now become critical, for the enactment of legislation to prevent from selling pesticides those who make recommendations for pest control or who are professional applicators. Enforcement of the regulations emanating from such legislation would do much to correct serious problems that continue to impede progress in pest management.

Lagging Research and Registration Difficulties

One of the basic principles in the concept of pest management is that appreciable levels of infestation by many pests can be tolerated by crop species without measurable effects on yield or quality. Furthermore, subeconomic levels of pest populations are often desirable to provide food for predators and parasites sufficient to build up their populations and hold them in the crop ecosystem. The ideal insecticide would be a strictly monotoxic compound to which the target species could not develop resistance and which could be produced cheaply.

The insecticide industry has not been able to produce such compounds. It is doubtful that it has much interest in doing so. Instead, it has concentrated on developing polytoxic chemicals effective for control of everything from *Abacarus hystrix* (Nalepa) to *Zygogramma exclamationis* (Fabricius). The high costs of research and development are claimed to have dictated such a policy.

Public criticism of insecticides during the last decade has been partially responsible for badly eroding research efforts devoted to discovering new and more effective conventional chemical insecticides. This situation has been worsened by the often frustratingly difficult problems encountered in obtaining clearance for use of new chemicals, or new uses of old chemicals. The result has been that research on conventional insecticides has continued to decline, in spite of the fact that conventional chemical insecticides are the backbone of insect control and will likely remain the backbone in the foreseeable future. Research on conventional pesticides should receive high priority.

Difficulties in registering for use some of the microbial insecticides needed for pest-management systems are even more frustrating and hard to understand. For example, the baculovirus of cabbage looper, *Trichoplusia ni* (Hubner) occurs in nature throughout the United States. It may persist in the soil for years in amounts sufficient to control cabbage-looper populations. Even so, until very recently there has been little encouragement given to attempts to register this useful pathogen.

It has been difficult to get across to regulatory officials, who are used to pesticides with mortality levels of ninety percent or more, that substantially lower levels of mortality are all that is required for microbial pathogens to be highly effective components of pest-management systems.

The necessary registration procedures must be simplified and accelerated. As we move toward increasing usage of pheromones, hormones, attractants, repellents, and chemosterilants, problems involved in registration are likely to become even more complex unless action is taken to develop efficient and reasonable registration procedures.

Adverse Effects of the Panacea Philosophy

Emphasis on the search for panaceas, of which the eradication concept is one of the best examples, has severely handicapped pest management. The biological activity of the synthetic organic insecticides introduced after World War II raised hopes for eradicating pests from the United States (Lyle 1947). The remarkable success of the sterile-male release technique in eradicating the screwworm from some areas and suppressing it to extremely low levels in others further excited the imagination of entomologists and stimulated much optimism for the eradication philosophy.

Although there have been rare examples of success, efforts to eradicate a species from large areas have most often failed. In spite of two recent large, expensive, and unsuccessful eradication attempts—against the imported fire ant and the screwworm—another program is being initiated that in both magnitude and cost will dwarf the other two combined.

An experiment designed to test the feasibility of eradicating the boll weevil from a large area in southern Mississippi was concluded during the summer of 1973. Eradication was not achieved. Nevertheless, another experiment in eradicating the boll weevil, involving 70-100 thousand acres in northeastern North Carolina and Virginia, is being planned for initiation during 1975. In June 1972 I made the following statement concerning the Mississippi experiment (Newsom 1974):

> We are now involved in a very large and expensive effort to eradicate the boll weevil from a large area in three states. I supported this undertaking with the understanding that it would be conducted for a two-year period after which it would be discontinued if eradication from the area was not achieved. I do not believe that it will be achieved. Based on past performance, I do not believe it will be discontinued if its objective is not achieved.

I should like to make a similar prediction for the proposed North Carolina and Virginia experiment. It will not succeed and the effort to eradicate the boll weevil will not be abandoned even if the experiment demonstrates that the technology for eradicating the boll weevil is not available.

The time is long overdue for the entomological profession to take a stand on the propriety of expending such a large percentage of the national effort and funds available for entomology on eradication programs. Overwhelming evidence suggests that the effort and funds could be put to far better use on pest-management research and implementation.

Prospects For Future Success

The problems listed above are serious impediments to the successful practice of pest management. They deserve careful consideration and prompt action. In spite of these obstacles and others that may arise, I predict that pest management will make the next twenty-five years the most exciting and productive period in the history of our profession.

The strategy of pest management has been proved effective for more than seventy-five years. Tactics that are considerably more effective than those available to the early workers are being made available. Even more importantly, entomologists are learning how to do the research necessary to make the practice of pest management successful. Research based on an interdisciplinary approach at the crop-ecosystem level, as in the NSF/EPA Integrated Pest Management Project, is highly effective and has generated much intellectual excitement. Modeling has proved to be of almost inestimable value in identifying needed research, handling the vast amounts of data being generated, and reducing to comprehensible dimensions the complex interactions characteristic of crop ecosystems.

There are already in operation, in various countries and at varying levels of development and sophistication, pest-management systems on row crops including cotton, soybean, sugarbeet, sugarcane, and tobacco; vegetable and greenhouse crops; fruit crops including apple, peach, and grape; coconut and oil palms; forest and ornamental trees. The numbers of these programs may be expected to increase rapidly. Pest management is like growing old, the alternatives are highly unsatisfactory.

REFERENCES

Adkisson, P.L. 1972. The integrated control of the insect pests of cotton. Proc. Tall Timbers Conf. Ecol Anim. Control by Habitat Manage. 4:175-88.

Brazzel, J.R. 1961. Destruction of diapause boll weevils as a means of boll weevil control. Texas Agr. Exp. Sta. Mimeo. Publ. 511.

Ewing, K.P., and C.R. Parencia, Jr. 1949. Experiments in

early season applications of insecticides for cotton insect control in Wharton County, Texas, during 1948. U.S. Dep. Agr. Bur. Entomol. and Plant Quar. E-772.

Falcon, L.A., and R.F. Smith. 1973. Guidelines for integrated control of cotton pests. Food and Agriculture Organization, United Nations, Rome.

Garman, H., and H.H. Jewett. 1914. The life history and habits of the corn earworm. Ky. Agr. Exp. Sta. Bul. 187.

Hunter, W.D., and W.E. Hinds. 1904. The Mexican cotton boll weevil. U.S. Dep. Agr. Bur. Entomol. Bul. 45.

Hunter, W.D., and W.D. Pierce. 1912. The Mexican cotton boll weevil, a summary of the investigations of this insect up to December 31, 1911. U.S. Senate Doc. 306.

Isely, D. 1924. The boll weevil problem in Arkansas. Ark. Agr. Exp. Sta. Bull. 190.

———. 1926. Early summer dispersion of the boll weevil. J. Econ. Entomol. 19:109-10.

———. 1934. Relationship between early varieties of cotton and boll weevil injury. J. Econ. Entomol. 27:762-66.

Lloyd, E.P., F.C. Tingle, J.R. McCoy, and T.B. Davich. 1966. The reproductive diapause approach to population control of the boll weevil. J. Econ. Entomol. 59:13-16.

Lyle, C. 1947. Achievements and possibilities in pest eradication. J. Econ. Entomol. 40:1-8.

Mitchell, E.B., and D.D. Hardee. 1974. In field traps: a new concept in survey and suppression of low populations of boll weevils. J. Econ. Entomol. 67:506-8.

National Academy of Sciences. 1969. Insect-pest management and control. Publication 1695. Washington.

Newsom, L.D. 1974. Pest management: history, current status, and future progress, p. 1-18. *In* Maxwell and Harris [eds.], Proceedings of the Summer Institute on Biological Control of Plant Insects and Diseases. University Press of Mississippi.

Pierce, W.D. 1922. How insects affect the cotton plant and means of controlling them. U.S. Dep. Agr. Farmer's Bull. 890.

Quaintance, A.L., and C.T. Brues. 1905. The cotton bollworm. U.S. Dep. Agr. Bull. 50.

Stern, V.M., R.F. Smith, R. van den Bosch, and K.S. Hagen. 1959. The integration of chemical and biological control of the spotted alfalfa aphid: the integrated control concept. Hilgardia 29:81-101.

Tumlinson, J.H., D.D. Hardee, R.C. Gueldner, A.C. Thompson, P.A. Hedin, and J.P. Minyard. 1969. Sex pheromones produced by boll weevil: isolation, identification, and synthesis. Science 166:1010-12.

SUBJECT INDEX

D

E

F

G

H

I

K

L

M

N

B C 7 D 8 E 9 F 0 G 1 H 2 I 3 J 4